The Game Changer

The Game Changer

The Next Generation Science Educators Today

Ovid K. Wong

ROWMAN & LITTLEFIELD
Lanham • Boulder • New York • London

Published by Rowman & Littlefield
An imprint of The Rowman & Littlefield Publishing Group, Inc.
4501 Forbes Boulevard, Suite 200, Lanham, Maryland 20706
www.rowman.com

86-90 Paul Street, London EC2A 4NE, United Kingdom

Copyright © 2023 by Ovid K. Wong

All rights reserved. No part of this book may be reproduced in any form or by any electronic or mechanical means, including information storage and retrieval systems, without written permission from the publisher, except by a reviewer who may quote passages in a review.

British Library Cataloguing in Publication Information Available

Library of Congress Cataloging-in-Publication Data Available

ISBN 9781475863437 (cloth) | ISBN 9781475863444 (pbk.)
| ISBN 9781475863451 (epub)

To the late Dr. Jack Easley, Jr., final examination committee chairman, and the late Dr. Charles M. Weller, director of Thesis Research at the University of Illinois, for their invaluable support. Thank you generously for sharing your wisdom and friendship. This book is also dedicated to the game changers to carry the torch of science education onward and forward for the many future generations to come with best wishes.

Contents

List of Figures and Tables	xi
Preface	xiii

1 Looking Back and Moving Forward in History — 1
- Anticipatory Questions — 1
- Sputnik Crisis 1957–1976 (Chaturvedi, 2020) — 2
- Nation at Risk (1983) — 5
- Project 2061 (1985) — 9
- Math–Science Partnership (2002) — 12
- Tapping America's Potential (2005) — 13
- American Competitive Initiative (2006) — 15
- A Nation Accountable (2008) — 16
- America Competes Reauthorization Act (2010) — 19
- The MAST Program (2011) — 20
- Academic, Social, and Emotional Learning Act (2011) — 21
- Next Generation Science Standards (2013) — 22
- Chapter Takeaway — 25

2 Auditing the Science Education Initiatives — 29
- Anticipatory Questions — 29
- The Qualitative Investigation (NSF-C-7621134) — 33
- The Quantitative Investigation (NSF DRL-1008228) — 41
- The National Assessment of Education Progress (NAEP 2019) — 49
- The Program for International Student Assessment (PISA 2018) — 52
- Trends in International Mathematics and Science Study (TIMSS 2019) — 54

viii   Contents

    What Is the Lesson Learned from the Triangulation of NAEP,
        PISA, and TIMSS?   55
    Chapter Takeaway   56

3 Executing the Intended Purpose of a Science Education Policy   59
    Anticipatory Questions   59
    Scenario I   59
    Scenario II   60
    A Communication Paradigm of Encoding and Decoding   61
    The US Department of Education (USED)   63
    The State Board of Education (SBOE)   64
    The Regional Office of Education (ROE)   67
    The School Board of Education   67
    The Science Classroom   68
    Chapter Takeaway   68

**4 The Intellectual and Social-Emotional Foundation of 2+2 = 4
versus 2+2 = 22**   71
    Anticipatory Questions   71
    Are You a Teacher-Centered Essentialist?   74
    Are You a Teacher-Centered Perennialist?   75
    Are You a Student-Centered Progressivist?   76
    Are You a Student-Centered Social Reconstructionist?   77
    Are You a Student-Centered Existentialist?   78
    Decisions, Decisions . . .   79
    The Social-Emotional Foundation of 2+2=4 versus 2+2=22   83
    What Is the Social-Emotional Learning (SEL) Baseline in Your
        Classroom?   87
    How Can You Be the Champion of Social-Emotional Learning
        (SEL)?   89
    Is Social-Emotional Learning (SEL) an Option?   96
    Chapter Takeaway   98

5 What Is the Contemporary Landscape of Science Education?   101
    Anticipatory Questions   101
    The Timeline of Major Science Discoveries   102
    Science Is Discoveries   104
    Science Is Technology   106
    Science and Its Major Divisions   109
    Physical Science (Chemistry)   112
    Physical Science (Physics)   115

	Life Science (Biology)	117
	Earth and Space Science	122
	Chapter Takeaway	127
6	How Do You Fill the Cup of Science Learning?	129
	Anticipatory Questions	129
	The Flyover	130
	Conceptual Learning and Teaching	130
	Learning Outcome	136
	Learning Background Check	139
	The Learning Experience	148
	Explore Learning	152
	Chapter Takeaway	156
Acronyms		159
About the Author		161

List of Figures and Tables

FIGURES

Figure 1.1	A 1957–2020 Science Education Timeline in the United States	5
Figure 2.1	The Science Education Initiative Audits	32
Figure 2.2	The Layered Issues of Science Education	40
Figure 2.3	The Science Teaching Variables	43
Figure 2.4	The NAEP, PISA, and TIMSS Triangulation	50
Figure 3.1	A Communication Paradigm of Encoding and Decoding	62
Figure 4.1	The Educational Philosophy Continuum	82
Figure 4.2	The Social Emotional Skills Matrix	98
Figure 5.1	The Scientific Method	110
Figure 5.2	Classification of All Things	112
Figure 5.3	The Periodic Table of Elements	115
Figure 5.4	The Big Picture of Life Science	118
Figure 5.5	The Big Picture of Earth and Space Science	123
Figure 6.1	Three States of Matter	133
Figure 6.2	Steps of Using a Light Microscope	133
Figure 6.3	The Rock Cycle	135
Figure 6.4	The Malawi Pregnancy Study	136
Figure 6.5	Fermilab Science Education Center	151
Figure 6.6	Fermilab Science Education Center	152
Figure 6.7	Plant Growth Experiment Simulation	153
Figure 6.8	Plant Growth Experiment Data	153
Figure 6.9	The Science Lesson with Technology Support	155

TABLES

Table 2.1	Average Science NAEP Science Scores 2015 and 2019	51
Table 2.2	PISA 2018 Science Literacy Assessment	53
Table 3.1	Science Themes of the Elementary Grades	69
Table 4.1	The Educational Philosophy Score Table	81
Table 4.2	Percentage of Positive Responses to Social-Emotional Questionnaires by Grade Level	86
Table 4.3	Classroom Social-Emotionality Teacher Self-Assessment Inventory	90
Table 5.1	Major Science Discoveries in the Seventeenth Century	102
Table 5.2	Major Scientific Discoveries in the Eighteenth Century	102
Table 5.3	Major Scientific Discoveries in the Nineteenth Century	103
Table 5.4	Major Scientific Discoveries in the Twentieth Century	103
Table 5.5	Major Scientific Discoveries in the Twenty-First Century	103
Table 6.1	Discrepant Events and Science Concepts	145

Preface

Public education is an intricate system of people engaging in learning and teaching. The system, regardless of its organizational complexity, follows rules and regulations of the local school district and state. Through interactive experience, stakeholders such as students and teachers follow set rules to achieve goals and objectives. How different is the education system from another one that also has interactive people following rules to achieve goals and objectives that we call it a game? How do the players engage to win if education is a game? Similarly, how do the players engage to win if education is not a game?

Some years ago, a young child asked his grandfather, "How do you play chess?" He was obviously intrigued by how two players can have fun over a chessboard engaged for hours. One simple way to have a game such as chess is to know that the game has players and that they follow the rules to play. The objective of chess is for the winner to capture the opponent's King so that the opponent cannot block the offense on his next move rendering the King helpless, or a checkmate. After years of playing my grandson Jayden became a skilled chess player, and he appreciated why chess is one of the world's most popular games, played by millions worldwide at home, in clubs, online, by correspondence, and in tournaments.

The young child later extended his interest to play other games and Risk is one of his favorites. Risk is a popular game using diplomacy, conflict, and conquest strategies with up to six players. The game board has a political map of the world, divided into territories grouped into continents. Dice rolls rotate the players, as they attempt to conquer territories with their soldiers. One unique feature of the game is that players may form and dissolve alliances with other players in the course of the game. The objective of the game is for the winner to take over as many territories as possible by eliminating other players. The easy-to-follow game rules and the fun interactions make Risk broadly entertaining to children, adults, and families.

Chess and Risk are two popular games, and people play these games because they want to win. Let us now explore one other example and find if it can be defined by what makes the interactive experience a game or not a game. The Stanley Parable is what we choose to further the discussion. The Stanley Parable is a computer video experience featuring Stanley, an isolated office worker. Stanley's main responsibility is to monitor computer information and follow through with no questions asked. One day, the computer went down leaving Stanley (i.e., the player) with no choice but to explore the surrounding elements of his office. When the player comes to a situation where a choice is possible, he can choose to follow or not follow the given directions by pressing buttons, or opening doors. Subsequently, the player is led to a number of predestinations that are all choice-driven.

The Stanley Parable has no hard and fast rules to follow leading the player to wonder about the purpose of the experience and whether it is a game as perceived in the traditional sense. Some players feel that the design of Stanley Parable is to explore the purpose of life and to achieve happiness through the pursuit of freedom. If Stanley Parable is a computer-generated game of life then can one question whether real life itself is a game? On the other hand, if the Stanley Parable experience is not a game, then can one ask what can be defined as a game?

What makes an interactive experience a game is a question we hear often in the real world of work, play, or home. When was the last time you heard the question, "Are you playing a game?" Here, we want to assume that the interactive experience enriches our lives in a way that we learn, enjoy, and improve and the list continues. We want to further assume that it is also the experience that is positive so as to benefit the common good. If the experience broadens the life's horizon, let it, and whether it is a game is honestly not important.

Do you want to limit your experience, or other people's experience assuming that you have the authority to do so? In that case, you call the experience a game and expect the players to follow the rules and not to explore beyond the predetermined boundaries of their job description. In the real work world, it is important to define people who play by the rules, the gamers, and those who do not play by the rules, the non-gamers. Why? In any workplace, we need gamers, who follow rules to support and stabilize the organization. Contrarily, we also need non-gamers who do not just follow the rules but create new rules to push the organization to reach new heights of excellence.

In many different interpretations of a non-gamer, the person is assumed to be revolutionary for whatever it is worth. The game changer crosses the "same old same old" gap to make things different. The *Jurassic Park* movie is a good example of a game changer. The original movie in 1993 made the next generation leap on the evolution of the Hollywood science action blockbuster. The 1993 movie debut probes into the intriguing concept of man versus dinosaur appealing to moviegoers of all ages. Since 1993, there have been six more in the mind baffling and visual dazzling movie sequel over twenty-nine years. The 2022 *Jurassic World: Dominion* is supposedly the last one in the sequel. Do you have doubts that Jurassic Park is not a game changer in movie making?

Are you now not convinced that there is a benefit having both the gamers and the non-gamers? Subsequently, in the big frame of reference, is the distinction between gamers and non-gamers an artificial division for people who are afraid to change, or justify their viewpoints?

Slowly but surely we realize that to define any interactive experience as a game is nothing more than putting a choke hold to moving forward in the name of progress. Similarly, rigidly defining many valuable life experiences such as critical thinking, problem solving as a game is pursuing a circular intellectual discussion similar to a dog chasing its tail.

Let us now transition the question of what is or is not a game to the critical life's experience of public schooling. "Is public schooling a game?" Schooling is a system with interactive components and for the most part between students and teachers. We all know that schools have rules. They have quantifiable outcomes regarding course grades and meeting the promotion requirements. Many people agree that public schooling in the United States has been less than successful because many students are not competitive and productive. A very important question that one should ask at this juncture is whether we need to uphold the status quo of schooling, continue to play the game, and follow the rules, or do we need to have the unorthodox intervention of a game changer.

There is no single one thing or process that is a game changer. If there was, we would have figured it out and adopted it by now. Do we need to look at all the new trends in learning and teaching? Absolutely. Keep in mind that our mindset toward successful schooling has to be open to many approaches, not just any specific one. The progressive mindset of schooling has to be inclusive of many research-proven approaches. To be successful, we need to hybridize the minds of the gamer and the non-gamer of education. The game changer which can be a policy, instructional materials, and technology, or the teacher himself is, and always will be, opening to new opportunities with purposeful human connection to learning and teaching.

From the broad spectrum of education let us now focus on the important area of science education. It is a very important discipline due to its relevance to students' lives and the commonly applicable skills of problem solving and critical thinking. More importantly, science education is one unique area that one can show the magnificent collaboration and dedication of professionals across the science disciplines with the groundbreaking game changers altering the inertia. If it is not for the collaboration of innovators to build from the basic science principles to the marvelous applications in technology and engineering, we will not have the convenience of the cell phone, the air fryer in the kitchen, the exploring rovers on the red planet Mars, and the mRNA vaccine to deter the pandemic and stop COVID-19 on its raging path.

On a personal note, the author of the book has gone through a rewarding science education career for almost five decades serving different organizations and institutions in various capacities. Five decades is a long enough time span for more than one generation in human chronology to witness the rhythmical pattern of coming and going. There are times when the professional game rules are changed and the players are to adjust their practices accordingly to survive and thrive. Allow me to look back standing on a hilltop to enlighten the next generation science educators on what I saw and learned to help you move forward.

The Game Changer: The Next Generation Science Educators Today challenges you to go beyond the shell of your educational practices. Do not be complacent and turn yourself to be the archivist of what you do. Be a reflective practitioner and seek new ways to make what you do more purposeful and effective. Be a game changer!

Ovid K. Wong, professor, School of Education,
Benedictine University, Lisle, Illinois

Chapter 1

Looking Back and Moving Forward in History

It is not often that nations learn from the past, even rarer that they draw the correct conclusions from it.

Henry Kissinger, US Secretary of State,
1973–1977, Nobel Peace Prize Recipient, 1973

ANTICIPATORY QUESTIONS

(1) What are the major national science education initiatives in the past fifty years?
(2) What triggered the national science education initiatives?
(3) How do you assess the success of the national science education initiatives?

History is the study of past events connected to human experiences. It helps us understand the significance of human experiences from a traditional perspective. Historians learn how people approached issues back in time on the one hand, and what could impact people moving forward in time on the other. Is this not the life's philosophy of looking back and moving forward?

History does not repeat itself exactly the same way because past events and issues are likely to reappear differently. Here is a reason we have the adage, "Study the Past" and "What is Past is Prologue." If we want to know how and why the world or parts of our world are the way it is today, we need to look to the past for clues.

According to many historians, a good way to study history is to cross reference a timeline of events that depict the evolution of happenings to include the who, the what, the when, and the why. History is the study of change, and understanding the role of change helps us interpret the mechanisms driving the change and its significance. There are many branches of history, and the heart of this chapter is the history of science education in the United States.

The history of the United States stands less than 300 years, and the tax-supported public education started early in the colonial period before 1776 in Boston, Massachusetts. More importantly, the recent sixty-five years are momentous in making the history of science education. Sixty-five years are a comparatively short span of time in the traditional sense of history, yet it is long in achieving global success in science.

Nobel prize winners for science in the fields of physics, chemistry, medicine, or physiology have been awarded to scientists who make the most outstanding contributions to mankind. For that reason, the Nobel award is regarded to be the most prestigious recognition worldwide. Since the first Nobel prizes were conferred in 1901, American scientists have won a whopping 269 science medals. How is that as an assessment of science achievement?

When is the beginning of science education in the United States? Depending on who you direct the question you may get two different answers. The education generalist will say "It started more than 300 years ago when people started to formalize education with public support." The science education specialist will answer "It started sixty-five years ago when the government started to pour funding to reinvigorate science and mathematics education."

During Colonial America in the 1600s and most of the 1700s, education was very different from the way it is today. The main school subjects were reading, writing, spelling, and religion. Mathematics, science, and other subjects were added only later. When education generalists use Colonial America as the beginning, it can be justified only as the predecessor of science and math education. On the other hand, the science education specialist will say 1957 is the beginning since science and mathematics education were given a major overhaul so students were prepared to be competitive with the former Soviet Union formally known as the Union of Soviet Socialist Republics or USSR for short.

In this chapter, 1957 is held as the spark to ignite the flame of science education revolution. Please note that science and mathematics education are commonly mentioned together because the two disciplines are highly integrated. We will visit the significant historical events in the United States to see how they change the progression leading to where we stand in science education today.

SPUTNIK CRISIS 1957–1976 (CHATURVEDI, 2020)

Many people fondly connect the book they love reading, the movie they enjoy, and the music they like listening to with a certain period of time. What year can you associate with the *New York Times* fiction best seller *Peyton Place* written by Grace Metalious, the Academy Award-winning best movie picture *Around the World in 80 Days*, or the top-list number-one hit song "All Shook Up" by Elvis Presley? Or are you old enough to remember when Dwight Eisenhower was the 34th President of the United States in 1957? More important than the book, the movie, and the music was the dawn of the golden age of science education. Let us find out.

The steady beep-beep-beep radio space signal was picked up by the United States several times a day when the Soviet artificial satellite Sputnik orbited the globe on October 4, 1957. Sputnik's radio signal pointed poignantly that the Soviet Union had surprisingly beaten the United States in space. Sputnik was the first man-made object ever to leave the Earth's atmosphere.

The Americans believed that the Soviets had sophisticated rocket technology capable of launching possible nuclear attacks in addition to another perception that there was an eye in space capable of spying on the United States. Under the administration of President Dwight Eisenhower, the event pointed at a threat to national security only short of triggering a potential military confrontation.

How did the United States respond to the triumphant Sputnik flight? The nation in essence met the challenge with resourcefulness and vigor through a bigger national defense budget, the formation of new national committees, policies, and an impactful overhaul of the educational system. But why education? The reason being that research, funding, and education were frequently used as the agenda of comparison between the rivaling Soviet Union and the United States.

In the wake of Sputnik, the Advanced Research Projects Agency (ARPA) was quickly established within the Department of Defense (DoD). ARPA was established to avoid unnecessary technological surprises like Sputnik. The charge of ARPA was to provide cutting-edge research that had the potential for significant technological achievement. The recurring question begged research to build on a strong foundation of science and engineering education.

A year after the Sputnik, the National Defense Education Act (NDEA) was established to empower a new generation of students to pursue college degrees in sciences and engineering. The undertaking focused its policy on science, mathematics, and foreign language instruction at all levels, from elementary through graduate school. NDEA galvanized the building of a strong science education foundation.

With the United States engaged in the Cold War with the Soviet Union, the pressure to prove the scientific and technological leadership of the United States was a national top priority. As an alternative to investing indiscriminately on research and development to any specific industrial objectives, the United States started up National Aeronautics and Space Administration (NASA).

In general, NASA is directed at generating cutting-edge knowledge and technologies to support the core missions of federal agencies. NASA was, and still is, a civilian agency responsible for peacetime aerospace research which was founded in 1958. NASA later launched numerous successful aerospace missions to the Moon and Mars to demonstrate that Americans are very competitive in the space program. It is apparent that for NASA to be successful, its staff members are to be accomplished professionals of science and engineering education.

Frequently, adequate funding determines the success of any operational initiatives. For that reason, the inclusion of the National Science Foundation (NSF) with reference to the current discussion of the post-Sputnik advancement of science education cannot be casually ignored. NSF created in 1950 is an independent agency of the United States government that supports fundamental research in science and engineering.

What is worth knowing is that in just two short years after the Sputnik, the NSF budget increased from 34 million dollars to 134 million dollars (Chaturvedi, 2020). It is obvious that the fourfold monetary increase said volume about the priority of science and engineering education. In the next chapter, we will study and analyze in-depth a landmark NSF-funded science education research projects to understand the down to the classroom impact of the science education reform.

The science education reforms after the Sputnik wakeup call were placed heavily in the hands of scientists focusing on upgrading the academic content. Many concerned educators argued that science education reform should be about what to teach, the academic knowledge, and how to teach, the pedagogy, and not one without the other.

During this period of reform, many national science programs were started to encourage and enhance science education in the United States including the Biological Science Curriculum Study (BSCS), Earth Sciences Curriculum Project (ESCP), Introductory Physical Science (IPS), Chemical Education Materials Study (CEMS), Intermediate Science Curriculum Study (ISCS), and Physical Science Study Committee (PSSC). During that time, many teachers fondly called the barrage of science curricula as the science alphabet soup. In a broad scope, the noted changes in reformed science curricula were rigorous in content supported by hands-on laboratory experiences.

Even though Sputnik was a modest artificial space satellite compared to the more sophisticated space devices to follow, the threatening beeping signal from space spurred the United States to pass significant education reforms known as the golden age of science education from 1957 to 1976 to reclaim science superiority it appeared to have lost to its Soviet rival. The Sputnik crisis and the subsequent education reform reinforce a very important science principle of cause and effect, similar to stimulus and response, that students from many science and mathematics disciplines are still learning today.

NATION AT RISK (1983)

(https://web.archive.org/web/20201029222248/https://www2.ed.gov/pubs/NatAtRisk/index.html)

What is your experience associated with the *New York Times* fiction best seller *The Little Drummer Girl* written by John le Carre, the Academy Award-winning best movie *Gandhi*, or two top-list number-one hit songs "Every Breath You Take" by the Police and "Billie Jean" by Michael Jackson? Or do you remember 1983 when Ronald Reagan was the 40th president of the United States? More important than the book, the movie, and the music culture was the 1983 grave Nation at Risk open report card to the American people. How do we shift from the Sputnik crisis to a Nation at Risk in twenty-six years? Let us find out.

Americans believe that all people are entitled to develop the potential of their mind to achieve the qualities needed to obtain gainful employment, manage their lives, and finally serve not only their interests but also the general well-being of the society. Americans also believe that shared education is critical to a democratic society and to the nurturing of a shared culture in a country that prides itself equally on pluralism and individualism. In essence, this is the philosophical ideal of the American public education. A Nation at Risk was the 1983 report of President Ronald Reagan's National Commission on Excellence. The front page report claimed the nation was at risk to

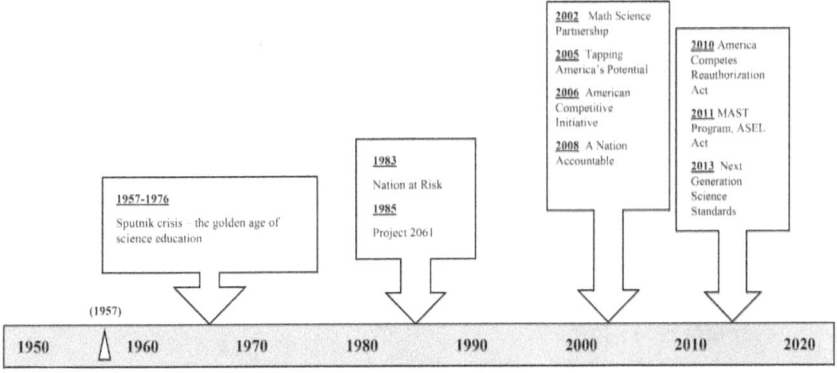

Figure 1.1 A 1957–2020 Science Education Timeline in the United States. *Source*: Author created.

reach the aforementioned ideal and called for stronger federal support for education.

A Nation at Risk is considered a modern American educational history landmark and the father of contemporary education reform. A quarter century had lapsed so what have the Americans not learned from the Sputnik crisis? Did we waste the gains in education made in 1957 and dissolve the support which helped make those gains possible?

The opening of the Nation at Risk Report says:

> Our Nation is at risk. Our once unchallenged preeminence in commerce, industry, science, and technological innovation is being overtaken by competitors throughout the world. This report is concerned with only one of the many causes and dimensions of the problem, but it is the one that undergirds American prosperity, security, and civility. We report to the American people that while we can take justifiable pride in what our schools and colleges have historically accomplished and contributed to the United States and the well-being of its people, the educational foundations of our society are presently being eroded by a rising tide of mediocrity that threatens our very future as a Nation and a people. What was unimaginable a generation ago has begun to occur—others are matching and surpassing our educational attainments.

"*A rising tide of mediocrity* . . ." is a capturing phrase used in the opening paragraph of the report, but what is the meaning of mediocrity in education? At heart, it implies that American students fell behind academically, that schools were failing because of inconsistent standards, and that teachers were inadequately prepared. The report remarked that the national security and the economy would collapse if the slackness in education continued.

What was the base for the Commission to determine that the country was at risk? The following learners' performance outcome indicators were cited.

(1) Low student achievement, especially minority student achievement, in comparison to other industrialized nations around the globe. The Scholastic Aptitude Test (SAT) in reading comprehension and mathematics was one of the assessment instruments used for the study.
(2) Declining science achievement as measured by national tests.
(3) Increased remedial mathematics courses offered at the college level. The increase in remedial courses of any discipline points to the inadequate readiness of the learners at a specific performance level.
(4) Concerned level of adult illiteracy required for most jobs and many everyday activities.

In sum, American schools did an inadequate job to prepare students for the real work world. With special mention to science education, it was regrettably noted that the United States was raising a new generation of Americans that was scientifically and technologically illiterate with a widening chasm between a small group of scientific elites and another group of scientific uninformed. The bottom line is that an increasing number of high school graduates were not prepared for college or for work.

How could the problems be resolved after the problems were identified? The following committee recommendations were based on the belief that students can learn to their fullest potential according to their abilities, career aspiration, and social economic background making a robust high school education essentially attainable to all.

In essence, the recommendations resonated with the get-tough rhetoric of the conformist movement. The following recommendations were made across the categories of content, standards, time, teaching, leadership, and fiscal support to meet the learning needs of both the college-bound and work-bound students.

1. Content

    Content includes the teaching of English, mathematics, science, social studies, computer science, and foreign language.

    The teaching of high school English should provide the learning of reading comprehension; well-written papers; effective listening and verbal skills; and literary heritage to understand the customs, ideas, and values of today's life and culture.

    The teaching of high school mathematics should provide the learning of geometric and algebraic concepts; fundamental probability and statistics; applied mathematics in everyday life; and the skills of estimation, approximation, and measurement.

    The teaching of high school science should provide the conceptual learning of physical and biological sciences; the processes of reasoning and critical thinking; and the connection of science to everyday life to promote the wellness of the society and the environment.

    The teaching of high school social studies should provide the learning of identity in the larger social and cultural structure; ancient and contemporary ideas that have shaped our world; the functions of the economic system and the political system; and the difference between free and oppressive societies. The learning of these areas is prerequisite to the informed and committed citizens in our free society.

    The teaching of high school computer science should be the learning of the computer as an information, computation, and communication device; the computer application for personal and work-related purposes; and the computer world and other related technologies.

The teaching of a foreign language should start in the elementary grades. It is desirable that students achieve such proficiency because the study of a foreign language introduces students to non-English-speaking cultures, heightens awareness and comprehension of one's native tongue, and serves the nation's needs in commerce, diplomacy, defense, and education.

2. Standards

The committee suggested that K–12 learning institutions accept more rigorous and measurable standards and set higher expectations of learning for colleges and universities to raise their admission requirements.

Grades should be used to indicate achievement and student readiness for the next level of study. Standardized achievement tests should be administered at key schooling transition points such as from high school to college or work. Here standardized achievement testing is not to be confused with aptitude testing. The purposes of the achievement tests serve to verify the student's qualifications, identify the need for remedial intervention, and recognize the opportunity for accelerated work. The testing is recommended as part of the State and local standardized assessment tests. In addition, diagnostic testing was also recommended to evaluate student progress.

3. Time

The committee recommended extended learning and teaching time of a 7-hour school day and a 200-day school year. For comparison, a typical school day runs from 8:00 a.m. to 2:30 p.m. and a regular school year runs 180 days.

4. Teaching

Improve the quality of teacher preparation and make teaching a more rewarding profession. Salaries for the teaching profession should be professionally competitive and performance-based. Furthermore, retention, promotion, and tenure decisions should be attached to a purposeful evaluation system of performance. Candidates preparing to enter the teaching profession should be required to meet the professional teaching standards, demonstrate the capacity to teach, and demonstrate competence in the academic discipline. Schools should cooperate to develop career ladders to distinguish the beginning teacher, the experienced teacher, and the master teacher.

To remedy the short supply of high school mathematics and science teachers, qualified individuals with mathematics and science degree with proper training, or retraining, could start teaching in these fields. The recommendation is pointing to the development of the alternative certification program for teaching degreed candidates with no formal training in education.

5. Leadership and fiscal support
    Elected officials such as the school board with the help of school administrators are the designated leaders needed to achieve these reforms. The legislators have the important role of setting meaningful policies and adequate fiscal support. Citizens as taxpayers are to provide the fiscal support and stability required to bring the reform efforts to a success. Excellence in education is costly; however, mediocrity in education in the long run is even more costly.

The 1983 Nation at Risk is a lengthy report; nevertheless, the reader can peruse it like many other education reports to realize that there are always recommendations to go with the identified issues or deficits. The deficits of Nation at Risk are student learning for the most part, and the recommendations for improvement are about more content rigor, and more learning time, higher expectations of the delivery system to include policy and financial support. In essence, it is about more everything to build a stronger and more secure nation.

PROJECT 2061 (1985)

(http://www.project2061.org/publications/articles/2061/sfaasum.htm)
Remember the *New York Times* fiction best seller *Skeleton Crew* written by Stephen King, the Academy Award-winning best movie *Amadeus*, two top-list number-one hit songs "Carless Whisper" by Wham! and "Like a Virgin" by Madonna? Or do you remember 1985 when Ronald Reagan served as the 40th president of the United States? More specific to our science education history than the contemporary culture was the groundbreaking Project 2061 project—Science for all Americans.

The year 1985 was a once-in-a-lifetime year for many because the Halley comet flew by the vicinity of planet Earth that people could see with their naked eyes. The comet will not return for another 75 years until 2061. Halley comet is a fascinating astronomical phenomenon for us to learn the many laws and principles in physical and earth-space science. The coincidental space event in time inspired the project's name. It was recognized that the school children who would possibly live to see the return of the comet in 2061, an approximate human lifetime from 1985, would soon be starting off their formal schooling.

Project 2061 is a long-standing research and development project of the American Association for the Advancement of Science (AAAS). The initiative took a visionary approach when compared to the Sputnik and Nation at Risk projects. The report showed no apparent mention about learning or teaching deficits and went straight to the benchmarks or standards needed to educate all Americans to become literate in science, mathematics, and technology. The benchmarks describe what all students should know and be able to do by the end of grades 2, 5, 8, and 12.

Project 2061 is considered a major milestone in science education because of the common core learning benchmarks approach that has the great science and educational influence for many years to come. Welcome to the dawn of the benchmark-driven curriculum, instruction, and learning.

The gem of the project is the expertise and organized description of the requisite knowledge and skills for science curriculum development and science education pedagogy. The description provides valuable scientific views of the world concerning the important things to learn in science.

The project's credibility is renowned because of the esteemed reputation of AAAS. The association is the world's largest general scientific professional society, with over 120,000 members, and is the publisher of the distinguished scientific journal *Science*. AAAS members are established scientists and specialists in their respective field of expertise. The following describes the views of the world from the eyes of science.

Science gives us the intellectual ability to make the world more comprehensible and more interesting. Amusingly, Project 2061, Science for All Americans does not encourage all students need to learn detailed knowledge of the scientific disciplines. Is this way of thinking different from many traditional ways of trivia science facts learning? Instead, the report advocates the in-depth learning of concepts over the superficial learning of facts.

Let us explore the following general concepts of science and mathematics. Science is divided into physical, life, earth, and space sciences. Mathematics is divided into computation, estimation, algebra, geometry, and probability.

The basic science concepts related to matter and energy with emphasis on their important interactivities explain diverse natural phenomena from the biochemistry of the elemental nonliving molecules to the complex living cells.

The universe is made of materials from the small base elements to the large space objects. The behavior of the elements is evolutionary, and they are directed by a few general principles such as the universal gravitational force and the conservation of energy. Co-accidentally, the composition of and the space-orbiting behavior of the Halley comet would be a suitable earth-space science topic of investigation. Earth in the solar system is the third planet from the sun. It provides us with an ideal habitation we fondly call home. The general features of the atmosphere, the geosphere, the hydrosphere, and the biosphere have critically influenced how we live and how human history has unfolded.

The world we live in is composed of the living, or the biological environment, and the nonliving, or the physical environment. The diverse living organisms share the unifying similarity of cellular structure and functions. Biological evolution is a concept based on evidence to expound the similarity and diversity of life forms. It is a central organizing theme for all of biology. The biotic and the abiotic elements are interdependent for the survival of all species of organisms and allow the proper management of matter and energy through all the complex life cycles.

The human life cycle through all stages of development and maturation, emphasizing the factors that contribute to the birth of a healthy child, to the fullest development of human potential, and to improved life expectancy. The basic structure and functioning of the human body represent a complex system of cells and organs that serve the primary functions of getting energy from food, protection against injury, internal coordination, and reproduction so life may continue.

Mathematics is a quantitative discipline dealing with symbols and symbolic relationships. It emphasizes kinds, properties, and applications of numbers and shapes, graphic and algebraic ways of expressing relationships. It coordinates systems of relating numbers to geometry. Probability concerns the numerical descriptions of how likely an event is to occur, or how likely it is that a proposition is true. It deals with estimating and expressing methods in uncertain situations to predict outcomes. Data analysis in mathematics uses numerical and graphic ways of summarizing data, the nature, and the limitations of correlations. Last but not least, reasoning behind the mathematical mind is the conscious production of mental thought with the use of logic. It includes restrictions of deductive logic, the uses and risks of generalization from limited experiences, and reasoning.

Do you notice some differences now that a science view of the world is presented in Project 2061? Do you note the borders between traditional subject-matter classifications are softened and integrations are underscored? Energy transformations, for example, are evident across the physical, life, earth, and space sciences.

Do you also notice that the minute pieces of information that students are expected to retain are considerably less than in typical science and mathematics courses? Concept learning and critical thinking are emphasized in lieu of vocabulary and factual regurgitation. The concept sets are selected to provide a stronger foundation of learning. Details are treated only as a way of enhancing students' understanding of general ideas. Any discipline of science should be learned and not memorized because the discipline is more than just a long list of vocabularies. The big ideas are the strong anchors of learning and not the loosely connected individual weak facts and words.

Lastly, for Project 2061, the funding makes it possible to create new informational booklets, professional development workshops, and electronic tools to meet the needs of a wide range of formal and informal science educators and members of the public. The funding is honestly nothing compared to the funding during the Sputnik crisis years.

MATH–SCIENCE PARTNERSHIP (2002)

(https://www2.ed.gov/programs/mathsci/msppp08.pdf)

The year 2002 is the new beginning of the 2000s decade, the second millennium. Many people are more familiar with the major events because the time is within the range of their lifespan. For that reason, many are likely to be familiar with the Academy Award-winning best movie "A Beautiful Mind." Chill, dude, and dawg were words thought to be hip in campus conversation. More important than the fad at the time was the silent run of another science education reform, the Mathematics and Science Partnerships (MSP) program by the Office of Academic Improvement under the US Department of Education.

The aim of MSP is to improve the content knowledge of teachers and the performance of students in science and mathematics. MSP encourage states, institutions of higher education (IHEs), local education agencies (LEAs), and elementary and secondary schools to participate in programs as partners. In this sense, MSP invites institutions and agencies to apply through non-competitive awards, based on a predetermined formula called a formula grant program. In solving problems in education, there is always more than one way to slice it.

The liberals tend to tackle the receiving end of education we call learning. On the other hand, the conservatives tend to look at education from the delivery end that we call teaching. Therefore, education improvement basically boils down to two areas—what to teach and how to teach—and MSP focuses more on what to teach.

Partnership is a unique feature of MSP. The partnership is an agreement by two or more parties to manage an operation and share the outcomes. To be eligible, an MSP partnership will include a minimum of a university science, mathematics, or engineering department, an IHE, and a high-need LEA. In education, a high-need LEA means schools with low social economic status and student achievement.

Conceptually, the following fours components describe the MSP framework and they are (1) form a partnership between IHE's science and mathematics faculty and high needs schools, (2) provide professional development to booster teachers' content knowledge, (3) improve classroom instruction, and (4) improve student achievement in science and mathematics. It is true that there is a ripple effect in all that we do that what teachers do touches the students.

How often have we heard teachers say, "If only I had time, I would write that wonderful idea up and use that in the classroom." MSP staff development projects, especially in the summer break, provide that key element of time. Continuous staff development in education is what teachers do to distinguish the profession from other trades or crafts.

Networking is a word in heavy use nowadays. In the real world of education, teaching is an isolated activity in a box we called the classroom where teachers work in isolation from their colleagues except for department meetings. There has not been enough time to talk about significant professional ideas in-depth. MSP teachers learned early that if they listen carefully, other people share the same concerns and seek the same solutions to similar problems in science and mathematics education. For that reason, a network of science educators has been growing thanks to the MSP initiative.

MSP is a continuous process of collaboration and improvement. As some problems are solved, new problems surface. Life goes on, and change is inevitable over time. MSP under the US Department of Education is not to be taken as an instamatic snap-shot but a full-length motion video.

MSP program indicators show that of the 57,000 educators served by MSP projects in 2008 (Abt Associates Inc., 2008), over two-thirds of these educators exhibited significant gains in their content knowledge (67 percent in mathematics and 73 percent in science). These educators, in turn, are enhancing the mathematics and science education of their students—over 2.8 million in 2008.

In 2012, 67 percent who were assessed in science showed significant gains in their content knowledge, and 63 percent of participants who were assessed in mathematics showed significant gains in their content knowledge. The proportion of students taught by MSP teachers who scored at the proficient level or above in state assessments of science remained strong in 2012, while in mathematics the proportion of students who scored proficient or above dropped. In science, the proportion of students scoring at the proficient level or above rose to 69 percent from 67 percent in the previous two years (US Department of Education. https://www2.ed.gov/programs/mathsci/mspperf-period2012.pdf).

TAPPING AMERICA'S POTENTIAL (2005)

(https://www.uschamber.com/sites/default/files/legacy/reports/050727_tap-statement.pdf)

The years 1997 to 2012 are the year range of Generation Z. It is the first social generation to have grown-ups with access to the Internet and portable digital technology from a young age; therefore, members of Generation Z have been dubbed "digital natives." In addition, Generation Z is the youngest, most ethnically diverse, and largest generation in American history, comprising about a quarter of the US population. Members of Generation Z spend more time on electronic devices, less time reading books, and this has profound implications for their attention span, what they learn, and thus their academic achievement as reflected in school grades. In 2005, George W. Bush was the 43rd president of the United States. The same year, the nation kicked off another science education innovation initiative: Tapping America's Potential (TAP).

How surprised are you when a science education reform proposal comes from the business community? What is the connection? It is only common for an employer to complain about a significant number of employees lacking basic skills in numeracy and literacy. A weakness in these basic skills can affect everyday performance as in the inability to extract information from printed texts and instructions, produce written reports, or work through calculations and make sense of numerical information.

In a similar vein, science education is perceived by the public, especially large industries, as dominating because it changes people's lives and the development of societies. Science knowledge with its applications is bringing immense transformations to business systems in an unparalleled way. TAP believes that America's wealth and resources in terms of the production and consumption of goods and services lie with its next generation of workers and what they can do to develop new technologies and products.

Mathematics and science education in the United States must be strengthened. The business organizations formed the TAP coalition to advocate for renewed attention to America's competitiveness and capacity for innovation. The fifteen prominent business representations are listed below.

The fifteen members are Aerospace Electronics Association, Business-Higher Education Forum, Business Roundtable, Council on Competitiveness, Computer Systems Policy Project, Information Technology Association of America, Information Technology Industry Council, Minority Business RoundTable, National Association of Manufacturers, National Defense Industrial Association, Semiconductor Industry Association, Software & Information Industry Association, TechNet, Telecommunication Industry Association, and US Chamber of Commerce. It was apparent that the business saw something that may escape the eyes of the public. They represent the receiving end of the education process, hiring the graduates as the workforce.

The ambitious TAP goal was based on the fierce international competition and the reliance on foreign talents. TAP was trying to build and sustain science, mathematics, and technology education as a top priority through public and private support. To sustain American competition, the business needs to inspire, recruit, and train a larger domestic pool of technical talent. This is so vital for the security and continued prosperity of our country that we can no longer delay action. A focused, long-term, comprehensive initiative by the public and private sectors called on business leaders to unite with government officials at all levels—national, state, and local—to create the momentum needed to achieve the priority goal.

TAP is perceived by science education historians as a boost to science education reform although there were no looming national security issues similar to the Sputnik era some fifty years ago. The TAP recommendations follow and they are:

(1) Build public support for making science, technology, engineering, and math improvement a national priority.
(2) Motivate US students and adults to study and enter science, technology, engineering, and mathematics careers, with a special effort geared to those in currently underrepresented groups.
(3) Upgrade K–12 math and science teaching to foster higher student achievement.
(4) Reform visa and immigration policies to enable the United States to attract and retain the sharpest and brightest science, technology, math, and engineering students from around the world to study for advanced degrees and stay to work in the United States.
(5) Boost and sustain funding for basic research, especially in the physical sciences and engineering.

AMERICAN COMPETITIVE INITIATIVE (2006)

(https://georgewbush-whitehouse.archives.gov/stateoftheunion/2006/aci/aci06-booklet.pdf)

The year 2006 was another eventful year in science and technology in the United States under the administration of President George W. Bush. NASA proudly revealed photographs taken by Mars Global Surveyor suggesting the presence of liquid water. Could Mars be the next interplanetary destination of the earthly migration in the future? The New Horizons probe was launched by NASA on the first mission to Pluto being reclassified by the International Astronomical Union (IAU) as a dwarf planet. What can one say about the advancement of science and technology when it took 9.5 years for the earthly probe to reach Pluto! In addition to all the mind-blowing deep space exploration, Twitter was launched in 2006 to become one of the largest social media platforms in the world.

President Bush in his 2006 address to the American people says:

One of the great engines of our growing economy is our Nation's capacity to innovate. Through America's investments in science and technology, we have revolutionized our economy and changed the world for the better. Groundbreaking ideas generated by innovative minds in the private and public sectors have paid enormous dividends—improving the lives and livelihoods of generations of Americans. To build on our successes and remain a leader in science and technology, I am pleased to announce the American Competitiveness Initiative. The American Competitiveness Initiative commits $5.9 billion in FY 2007 to increase investments in research and development, strengthen education, and encourage entrepreneurship.

The initial amount of $5.9 billion should not be taken lightly because in ten years the incessant investment will add up to more than $50 billion. What can the American people spend the dollars in training and innovations? In general, the support was designated for the science-math professionals, workers, teachers, and students to include new opportunities for the underprivileged with new Title I funding allotment for people on low incomes.

Although the American Competitiveness Initiative (ACI) proposed millions in new federal dollars to strengthen the nation's education system the focus support remained in the improvement of K–12 science, math, and technological education. Fundamental to improving student learning and achievement is the classroom presence of highly qualified teachers. Effective teachers must have mastery of content and pedagogy to engage students in rigorous courses that teach critical problem-solving skills to better prepare future citizens in the global marketplace.

While the federal programs are traditional in the recruitment of more college science and math majors into teaching, the required education training means that it takes time to fill classrooms with highly qualified science and math teachers. Interestingly, there is an untapped resource among current and retiring science and mathematics professionals who have the content mastery and the practical experience to serve as effective science or math teachers.

Is it possible that math and science professionals are more likely to transition to the teaching career if the teacher certification process recognizes the equivalent training and experience these professionals have in their field of expertise? Are we looking for a different way to train and recruit teachers? To meet these challenges and questions, ACI proposed a different approach that provides professional development for current teachers and attracts new teacher candidates to the classroom.

A NATION ACCOUNTABLE (2008)

(https://www2.ed.gov/rschstat/research/pubs/accountable/accountable.pdf)

There were a number of firsts in 2008. China hosted the 2008 Summer Olympics for the first time in Beijing. Many saw the big event as a soft-power showcase that would advance the country as an emerging global superpower. In the 2008 Olympics, remarkably, the American swimmer Michael Phelps was the first person ever to win eight gold medals in one Olympic game.

In the political arena, Barack Obama in 2008 was the first man of African American descent to be elected as the president of the United States. He was the 44th president of the United States serving from 2009 to 2017.

Almost five thousand miles away from America, Switzerland powered up the large hadron collider at Conseil European pour la Recherche Nucleaire (CERN) in Geneva. The collider is the world's largest of its kind, and it is also the highest energy particle collider. CERN has the responsibility to train the next generation scientists to bring nations closer together. CERN, Fermilab (Illinois, USA), and many other international institutions are all partners in this ambitious science and technology research endeavor.

Despite the general perceived global leadership of the United States, a Nation at Risk in 1983 after twenty-five years remains a Nation at Risk if not more at risk than in 2008. It is disheartening that the system had not been keeping up with the global economy, the demographic changes, and the education challenges that come with the shifts.

To be more explicit about the humiliation of the American school system, let us appraise a typical group of twenty children who started kindergarten as a 1988 cohort. In 2001, six of them would not have graduated on time. By 2007, of the fourteen who would have graduated on time, ten would start college, and only five of those twenty kindergartners would have a college degree (US Department of Education, National Center for Education Statistics Digest of Education Statistics 2007. Tables 102, 193, and 318. Washington, DC, 2008. College graduation rates for 2007 are estimated from the most recent available). The performance of the school system is in good conscience not acceptable, and it is in essence a paper tiger—a beast perceived to be strong but is in reality weak.

In 2001, Americans finally had the opportunity to take a comprehensive assessment of the school system thanks to the standards and accountability movement of the No Child Left Behind (NCLB) Act. This pivotal law, which was passed with bipartisan majorities in Congress and with the support of the business and civil rights communities, ensured that schools taking government support agree to measure and report on results with reference to learning standards and accountability.

NCLB mandated the education-outcome accountability and translated that to the no-nonsense annual public school report with student testing data. NCLB in reality is a hard-pressed school curriculum and instruction that are standard and accountability-driven.

Under NCLB, all classroom teachers are highly qualified to mean that the instructors need to meet a set number of college credits in their area of teaching. The number of teachers with less than the required college credits in the discipline are discouraged and reported in the school report card. A chemistry teacher, for example, will not be allowed to teach a different science course outside chemistry without the proper endorsement stated in the teaching license. Under NCLB, the school principal goes beyond the managerial responsibility to take on the new role of being the instructional leader. The school administrator is to spend more time in the classroom supervising teachers than staying in the office pushing paperwork.

Under the guise of NCLB, many teachers experienced the controversial common core standards (CCS) mandate which some view as an attack on states' rights on what to do in education tied to federal funding. However, others know the mandate as a progressive thought-out education reform that would tackle equity issues and improve education in a global society.

What are CCS (Bidwell, 2004)? They are indexes that define what students should know and be able to do that are subject matter and grade level specific. In many states, the State Board of Education decides on the learning standards for all students, from kindergarten through high school. The State Department of Education helps schools make sure that all students meet the standards, and this is done by the administration of the annual state assessment system.

Subsequent to the CCS mandate, a number of states across the nation have adopted the same standards for English and math. These standards are called the Common Core State Standards (CCSS). Having the same standards helps all students to get a good education, even if they change schools or move to a different state. Teachers, parents, and education experts designed the CCSS to prepare students for success in college and the workplace. The CCS has a lasting impact in schools, and the buzzwords in education nowadays remain a standard-driven curriculum, instruction, and assessment.

To date, all fifty states have reading and math content standards and tests at a minimum in grades 3 to 8 and once in high school. The annual school report card shows test results of all student subgroups by race, English language learners, students with learning disabilities, and students from low-income families that are now publicly available on the Internet. Publication of these test results means that taxpayers, educators, and policymakers now have more information at their disposal to enable a more purposeful national dialogue on education.

In the past, many educational reforms were the push for an increase in funding. Unfortunately, that support has not gone closely with student learning improvements. While resources will always remain a hot debate in education, there need to be other more effective and efficient ways to use funding support. On the one hand, the country continues to be a Nation at Risk, on the other hand, she was also a nation informed and accountable, and recognized that there still was much work ahead.

AMERICA COMPETES REAUTHORIZATION ACT (2010)

(https://www.congress.gov/111/plaws/publ358/PLAW-111publ358.pdf)

2010 was a year with many global natural disasters. The top headline news was the Haiti's 7.0 magnitude earthquake that killed over 300,000 people and wiped out the greater part of the capital city, Port-au-Prince. *Avatar* (a 2009 film) awed the viewers as an alien-jungle virtual-reality action-adventure 3D sensation that grossed $2 billion worldwide, a first in the list of highest-grossing films. 2010 was a heavily political year, as evidenced by the prevalence of high-ranking politicians, and President Obama was top-ranked as the most influential person of the year.

Reauthorization of the America Competes Act rejuvenated a vigorous investment in basic research and education focusing on science, technology, engineering, and mathematics or STEM as a national priority. It is only natural for the federal government that they continue the financial support to sustain the efforts to be a leader in innovation and ingenuity. The Act is about the development of new products manufactured in America, the application of sustainable energy, the improvement of better health care with the use of information technology, and the better protection for our citizens. In all these endeavors, innovation will be key to achieving the goals.

The Reauthorization Act of 2010 is the successor of the America Competes Act between 2008 and 2010. Subsequently, the Reauthorization Act is a second three-year period of funding for science and technology.

How does STEM work in a traditional classroom? For one important pointer, STEM is not the teaching of a single subject as one might see in a conventional school curriculum. It integrates key concepts between two or more STEM disciplines as students apply the practices of science and engineering to real-world problems. STEM projects integrate science, mathematics, and technology as in many problem-based robotics projects. Let us find out more.

The mousetrap car, for example, is popular and a classic challenge in innovative thinking. Students are to build a contraption that can travel by using only the energy that is stored in the spring of a single mousetrap. The success of the contraption design is evaluated by the speed and the distance traveled by the vehicle. The concept of the mousetrap car is surprisingly rich in science, applied math, and design. As a science teacher, it is inviting to use a STEM lesson to connect and integrate different disciplines with real-world applications in an enjoyable learning experience with the intent of training the next generation of innovative thinkers and doers.

The America Competes Reauthorization Act was the moment that America rebuilt a new and stronger foundation for economic growth and global leadership. Americans needed to do what America has always been known for in building, innovating, and making things, and we start doing all that by reinforcing the foundation of education.

THE MAST PROGRAM (2011)

While Barack Obama was serving out his second term of presidency, two farewells were recorded as good news stories in 2011. After many nuggets of wits and generous giveaways, Oprah Winfrey, the queen of daytime talk shows, ended her popular syndicated performance after 25 years on over 200 US television stations in more than 100 countries globally. The same year, Harry Potter after 450 million books ended the reign of the boy wizard in the movie *Harry Potter and the Deathly Hollow: Part 2* as the last installment in the film series. Oprah Winfrey and Harry Potter offer something interesting for everyone, which is why they reached a wide audience, and will continue to teach lessons to people coming in contact with the show, the book, and the media. As educators, do we not wish that what we do can have the same if not more positive impact like Oprah Winfrey and Harry Potter?

In hiring a public school teacher, the State Department of Education needs proof that the candidates are competent in teaching the subject matter with proper classroom management skills. Normally, an education degree is that proof. To that end, the two options to be a licensed teacher are the four-year traditional college degree program or the two-year alternative certification program. For many people who did not major in education when attending college, but instead worked in the private sector, and later wants to become a teacher, they now have the less time-consuming and less-costly route to be professionally licensed through the alternative teacher certification program.

Teachers are the key to student success, and science education is no exception to this generalization. Many IHEs due to the demand for science teacher shortage offer a nontraditional route for teacher training with specific reference to high school science teachers. The Master's degree programs usually take two years to complete, whereas certificate-only programs can be completed in more than one year in many states.

Due to the growing acceptance of the alternative routes, many colleges now offer online alternative teacher certification programs. MAST (Master of Arts in Science Teaching) is an alternative teacher training program designed to help science professionals become middle and high school teachers. In two years, these scientists can earn a Master of Arts degree with an emphasis on science teaching. This program was launched by the University of Nebraska-Lincoln.

Interestingly, during the career time of the author, he has worked with a fighter jet pilot with a BSE (Bachelor of Science in Engineering), a chiropractor with a DC (Doctor of Chiropractic), and a medical doctor with an MD (Doctor of Medicine). They all successfully completed the MAST program to become successful public school science teachers, and students are blessed to learn from these science professionals turned teachers.

ACADEMIC, SOCIAL, AND EMOTIONAL LEARNING ACT (2011)

(https://www.edutopia.org/social-emotional-learning-history)

Academic, Social, and Emotional Learning Act (ASEL) (2011) amends Title II of the Elementary and Secondary Education Act of 1965. It includes a teacher and principal training that addresses the social and emotional needs of students. Why is ASEL pivotal to the betterment of science education? Let us find out.

In a deceptive manner, many think that teaching is simple as filling an empty vessel or building a structure from the ground up. What science teachers need to realize is that effective teaching is not filling an empty vessel or building on an assumed foundation of mental readiness and receptivity of the students. Likewise, we can also say the same for the mental readiness affecting the effective instructional delivery of the teacher.

In a pathetic classroom one can describe that the student is mentally not ready to learn, or the teacher is mentally not prepared to teach, and worse yet both are emotionally not engaged. Honestly, how can a person do well if he is stressed and not paying attention? Social-emotional learning (SEL) is a very important variable in effective education now more than ever due to the crashes of the caustic societal unrest and the deadly pandemic.

Please note that a science teacher is effective not only because he is knowledgeable and skilled but also because he knows himself and his students well enough to stabilize the social-emotional aspect of the learning environment. Simply put, the teacher calms himself, and he also calms his student. In principle, social-emotional learning is an essential part of science education with reference to the communication process of learning and teaching.

The history of SEL is as old as ancient Greece. When Plato, the distinguished Athenian philosopher, wrote in *The Republic*, he proposed an all-round education approach that demands a balanced teaching in the arts, math, science, physical education, and character development. Fast forward to recent times, James Comer from the Yale School of Medicine's Child Study Center hypothesized that academic achievement of a child is affected by his home and school experiences. Do the experiences strike a chord of mental music or mental noise to learning?

As an example, the Illinois State Board of Education has three SEL goals, and they are goal 1: develop self-awareness and self-management skills to achieve school and life success; goal 2: use social awareness and interpersonal skills to establish and maintain positive relationships; and goal 3: demonstrate decision-making skills and responsible behaviors in personal, school, and community context.

How to achieve proper SEL in the classroom in addition to head knowledge education will remain an area of study and practice in progress. In education, the ability to manage your own emotions as a teacher as well as to recognize and interact with students and colleagues is an important skill to achieve success.

NEXT GENERATION SCIENCE STANDARDS (2013)

(https://www.nextgenscience.org/)

The dark side of 2013 was marred by the horrific bombing of the Boston Marathon with the killing of the mastermind and the capture of his accomplice. This tragedy was heart-tearing and a defining moment in the unlawful use of violence against civilians in the pursuit of political aims.

The same year, Elon Musk and Jeff Bezos capture the attention of the American public. Elon Musk is the confounder of Tesla, the maker of high-technology all-electric cars. He is the founder of Space X, a private spaceflight company that launches space satellites and delivers cargo and, more recently, crew to the International Space Station. If you are interested in space travel you might be interested to follow this entrepreneur in the news.

Jeff Bezos is another successful entrepreneur and the founder of Amazon, a worldwide e-commerce company. If you are an online consumer, are you not fascinated by the effectiveness and convenience of technology? The future world will be driven by innovative science applications; Elon Musk and Jeff Bezos are but two among other shining on the current world stage. Let us hope that more world-class inventors and entrepreneurs will come down the pike in the nearest future from your classroom!

Tracing the blazing career path of Elon Musk and Jeff Bezos, one can generalize that the two men initially follow the rules on fundamental matters to reflect that life in general will be in order with regulations. However, what made Elon Musk and Jeff Bezos eventually entrepreneurial is that they take the risk to break the rules in their career later to follow their life's passion and make a difference. They are the contemporary game changers of the time.

Is there a parallel between the career path of entrepreneurship and the academic path of education? What would learning and achievement be like if they follow a path with or without rules and regulations? The answer to the question is an appropriate transition to the presentation of standards and regulation in science education.

Before the final roll-out of the Next Generation Science Standards (NGSS) in 2013, American science education followed a guideline developed by a consortium of scientists, engineers, educators, and other field experts from twenty-six states to achieve the ultimate goal of scientific literacy. In one book definition, goals are broad directives for people to follow and complete tasks.

What follows is the pre-NGSS example of the 1995 Illinois science goals of (I) knowledge of science to include energy, stability and change, structure and function, systems and interactions; (II) science, technology, and society; (III) principles of scientific research and ethical practice; and (IV) the processes of science to include data collection, data analysis, graphing, units of measurement, and safety practices. The focus of the mandatory state K–12 science goals is reflected in the annual Illinois Goals and Assessment Program (IGAP).

In just a few years, the science education goals in Illinois, as an example, were fine-tuned to learning standards in 1999. The scientific inquiry and design goal are divided further to include processes of scientific inquiry and processes of technological design. The concepts and principles of science goal are further divided to include living things and their functions, living things and their environment, matter and energy, force and motion, earth features, processes, and resources, earth and the universe. The goal of science, technology, and society is further divided to include practices of science and interactions between science, technology, and society. The focus of the mandatory state K–12 science assessment is reflected in the annual Illinois Standards and Assessment Test (ISAT).

What does the change from the Illinois goals to Illinois standards tell us about the science education framework? What is the possibility that the standards are more manageable for learning/teaching, and they are also easier to measure and track student achievement?

The NGSS developed by scientists, engineers, and education experts is K–12 science for what students should know and be able to do with the intention of nationwide adoption. Twenty states and the District of Columbia have adopted the NGSS, and twenty-four states have developed their own standards based on the NGSS framework for K–12 science education. How can one not say that science education nowadays is not standard-driven?

The NGSS framework defines three dimensions that students must learn in unison to achieve scientific literacy. The dimensions are core ideas, practices, and cross-cutting concepts. One uniqueness of learning in unison to achieve a purpose is illustrated by the three-dimensional property of the NGSS Möbius strip logo. The ribbon-like strip is continuous though twisted and has several interesting properties that when a line is drawn along the paper edge travels in a full circle to a point opposite the starting point. In a way, the idea of unison is seen when the drawn line goes around in a full circle. Fascinatingly, the Möbius strip is studied to explore the thought-provoking properties of geometry and topology.

The three dimensions of NGSS are disciplinary core ideas, science practices, and cross-cutting concepts. We will delve into the details of the standards here because this will greatly enhance our understanding of the foundation of science education today.

What are the NGSS disciplinary core ideas? They represent a broad-based approach to understand the scope of science knowledge. The four groups are: (1) physical sciences, (2) life sciences, (3) earth and space sciences, and (4) engineering, technology and science applications. The details of each group follow.

(I) Physical Sciences (PS)
   PS1 Matter and its interactions
   PS2 Motion and stability (forces and interactions)
   PS3 Energy
   PS4 Waves and their applications

(II) Life Sciences (LS)
   LS1 From molecules to organism
   LS2 Ecosystems and interactions
   LS3 Heredity
   LS4 Biological evolution

(III) Earth and Space Sciences (ESS)
   ESS1 Earth's place in the universe
   ESS2 Earth's systems
   ESS3 Earth and human activity

(IV) Engineering, Technology, and Science Applications (ETS)
   ETS1 Engineering design
   ETS2 Connections among engineering, technology, science, and society

What are the NGSS science practices? In essence, they represent the science methods of investigation to include: (1) Asking questions; (2) Developing and using models; (3) Planning and carrying out investigations; (4) Analyzing and interpreting data; (5) Using mathematics and computational thinking; (6) Constructing explanations for science and designing solution for engineering; (7) Engaging argument from evidence; and (8) Obtaining, evaluating, and communicating information.

What are the NGSS cross-cutting concepts? In essence, the concepts represent seven big ideas across the science disciplines enabling integrated science learning. The concepts follow.

(1) Patterns; (2) Cause and effect; (3) Scale, proportion, and quantity; (4) Systems and system models; (5) Energy and matter; (6) Structure and function; and (7) Stability and change.

A flyover of the NGSS gives the off-the-cuff readers the impression that the standards are reasonably another version of the old science education framework. The view may be supported because the spirit of science knowledge and practices does not just die. Nevertheless, an in-depth examination of NGSS shows that what students need to know and be able to do in physical sciences, life sciences, earth and space sciences, technology and science applications are extended in addition to the cross-cutting concepts to cement all the areas. As a science educator, it will be almost impossible to provide a science learning experience beyond the four walls of the standards and that obviously speaks well for the inclusive design of NGSS.

CHAPTER TAKEAWAY

The evolution of science education in America since 1957 is best described as a combination movement of jumps, hops, and skips. To be clear, the three movements are diffcrent with reference to the intensity and the distance traveled. A jump is a quick and sudden move in a specific way. A hop is a jump by one foot, and the intensity is usually less than a jump. Finally, a skip is a light stepping from one foot to the other. The combination movements translate to moving in a direction with patterns of regularity.

B. F. Skinner, the Harvard psychologist, was right about the change in human behaviors that it does not occur in a vacuum. The evolution of science education over a span of sixty-five years is undoubtedly the jump start in 1957. The Russian Sputnik launch was the stimulus, and the American response was the catch up with the space race to reflect global competition and superiority.

After the initial jump start, Americans were reminded time and again that the initial efforts of improvement need to be sustained, or even amplified, to meet new challenges nationally and internationally. For many years, B. F. Skinner was right again in shaping human behavior through positive and negative reinforcements. The shaping of human behavior as in learning would be way too easy if the human minds are just empty vessels waiting to be filled.

Do you know that the challenge of behavioral modification for improvement needs a thorough systemic and systematic change with 77 million students, 3 million teachers, 98,000 public schools in 50 states? How is it possible that we regulate the huge US populace under a complex political system to move forward? Honestly, the task is daunting under the current system of education.

The 2008 American held accountable with the inherited 2001 NCLB Act is worth citing because the education system for the first time in a long time has consequential accountability penetrating its way down to the schools. A meager suggestion for improvement is more often than not to be taken seriously. On the other hand, when accountability is a critical part of the system, people will take it more seriously for either the reward of success or the consequence of failure. Education stakeholders need to be empowered to regulate healthy decisions and actions to prevent anyone from exerting too much power on self-interest to promote student learning and prosper the nation. Is this not what checks and balances are about?

Science education has trotted down a long and meandering path for more than sixty-five years since 1957. What was important ten years ago is not even meaningful today, but that does not mean that the many life's lessons pointing to the past are not important. We have the accumulated wisdom to deal with challenges of the future and the ability and flexibility to embrace all things as we move forward and onward.

REFERENCES

Abt Associates Inc. (2008). Mathematics and Science Partnerships: Summary of Performance Period 2008 Annual Reports. Retrieved on August 15, 2021 from https://www2.ed.gov/programs/mathsci/msppp08.pdf.

Academic, Social, and Emotional Learning Act. (2011). Retrieved on October 7, 2021 from https://www.edutopia.org/social-emotional-learning-history.

America Competes Reauthorization Act. (2010). Retrieved on October 13, 2021 from https://www.congress.gov/111/plaws/publ358/PLAW-111publ358.pdf.

American Competitive Initiative. (2006). Retrieved on September 25, 2021 from https://georgewbush-whitehouse.archives.gov/stateoftheunion/2006/aci/aci06-booklet.pdf.

Bidwell, A. (2004). The history of common core standards. Retrieved on April 4, 2021 from https://www.usnews.com/news/special-reports/articles/2014/02/27/the-history-of-common-core-state-standards.

Chaturvedi, P. (2020). Changes in the United States in response to the Sputnik crisis. Retrieved on May 30, 2021, from https://www.thetidingsblog.com/post/changes-in-the-united-states-in-response-to-the-sputnik-crisis.

A Nation Accountable. (2008). Retrieved on September 21, 2021 from https://www2.ed.gov/rschstat/research/pubs/accountable/accountable.

A Nation at Risk. (1983). Retrieved on August 5, 2021 from https://web.archive.org/web/20201029222248/https://www2.ed.gov/pubs/NatAtRisk/index.html.

Next Generation Science Standards. (2013). Retrieved on December 13, 2021 from https://www.nextgenscience.org/.

Project 2061. (1985). Retrieved on August 10, 2021 from http://www.project2061.org/publications.

Tapping America's Potential: The Education for Innovation Initiative. (2005). Retrieved on September 5, 2021 from https://www.uschamber.com/sites/default/files/legacy/reports/050727_tapstatement.pdf.

Chapter 2

Auditing the Science Education Initiatives

Audit findings are easy to come up with, successful change from a finding is the true internal audit value.

Michael Piazza, professional baseball catcher, MLB 1992–2007

ANTICIPATORY QUESTIONS

(1) What is the purpose of auditing the national science education initiatives?
(2) What is the difference between a qualitative and a quantitative study?
(3) How does triangulation help to make program assessment more credible?
(4) What is the current status of science education in the United States with reference to national and international science literacy?

Do you have a to-do list to guide you through the day? Check the same list at the end of the day, and you might surprise yourself by finding that the list is completed or just partially done. The next day you might hold on to the same list to check off the done items and add a few more new items. Would it not be nice for you to say one day that the to-do list is done and now you have some free time to yourself? We want to have a to-do list so our life's schedule is purpose-driven. We also want to check off the list to show that it is a task completed or mission accomplished.

In this chapter in a similar way, we want to be sure that the major science education initiatives are completed in an acceptable manner so we can continue to move forward to achieve new goals.

Once upon a time, there was an ancient kingdom where the people would change the ruler of the land every four years. The king had to agree to the term that he would go beyond his four-year contract only if he served his people well, or he would have to leave mid-term in two years if he did not serve his people.

One king finished his first term, and it was time for him to leave. The people put him in an arrogant chariot and paraded him around the kingdom to say goodbye to his people. This was the moment of sadness, and the king said "I thought I served my people well. Unfortunately, I did not get the confidence of my people and I had to leave."

The next king came along, and he was a wise man. He gathered around him an elite group of counselors advising him on matters related to finance, travel, communication, military, trade, health, and other services to the kingdom. He finished his first term successfully, and the people offered him a second term. The king said to himself "I served my people and I could just repeat what I did in my second term."

In the second term, the geopolitical climate changed suddenly. There were enduring foreign wars, and to make things worse there was a harsh lingering pandemic. People grumbled and complained about the king for using feeble old strategies to solve new problems. In the middle of the second term the people banished the king. Upon leaving the king said to himself: "My first term was a success and I had a group of advisors to support me. What happened?" Time passed and many rulers of the kingdom came and departed.

One year another young king came along. He was wise and flexible in adjusting his decisions and actions to meet the needs of his people. There was also a handpicked group of court counselors giving the king advice. The king soon realized that what he knew about his people was honestly limited. He said, "I would like to have in person experience with my people in lieu of hearing about them indirectly from far away."

After half a year of his reign, the king came up with the idea of visiting his people incognito. He assumed the identity of a farmer in the busy marketplace. His fresh and inexpensive fruits attracted many customers, and he had the opportunity to talk to the people about what was happening around town. In a few weeks, he was able to move his fruit stand to different parts of the kingdom. What was valuable to the king was his real-world experience of talking to the common people and learning about their lives under his rule. Is this not similar to auditing in today's business world?

After a month, the undercover king went back to his beautiful and comfortable castle. He met with his wise group and made in-depth studies about his findings. He compared the information that he gathered from his subjects and the decisions that he made for them. He adjusted his decision and action to the best of what he knew and to the rich resources of what his kingdom could provide.

At the end of the first term the people happily offered the young king to serve another term and he agreed. After several terms serving, people asked him "All the other kings served, and they were asked to leave before the term was up. What was your well-kept secret?" The wise king replied simply "I know my people, and I served their needs well." In the continuous passage of time, the wise king continued to rule merrily for many years. The monarch took action from the information that he gathered from the people and that was the leadership trait of the wise king.

In education research, it is common to use various methods to explore the real work world of learning and teaching. Is this approach not similar to the wise king visiting his kingdom in to study the lives of his people? The effective ruling of a system such as education may not be as complex as the ruling of the ancient kingdom that we just read. Nevertheless, what we learn from the story is that good rulings include making good decisions which are connected to good information. How then do we define good information?

One definition of good information is the in-depth collection of data from a scope of reliable sources. Social science researchers agree that the method used in a smaller number sample size allows an in-depth study of the problem in the qualitative descriptive term. The other definition of good information is the broad-scope collection of information from a large number of sources. The large sample size approach is good in the coverage of information. This method necessitates the delineation of the problem under investigation in quantitative measurable terms.

In the education community, researchers frequently debate the pros and cons of application of the qualitative and quantitative methods of investigation. Let us find out more.

Qualitative methods provide relevant information about reasons to answer the questions of how and why. This is where the qualitative methods help to study the natural setting, the behaviors, and the points of views of people. In education, our feelings, thoughts, and perceptions drive complex behaviors, and it is the qualitative methods that aim to investigate these psychological interactive elements of learning and teaching. The qualitative researchers look for what is common, pervasive, and dependable, allowing subsequent generalization bound by particular cases.

Quantitative methods, on the other hand, give measurable terms, such as "how long," "how many," and "how much." The information is then applied to various statistical methods for data analysis and interpretation. Qualitative versus quantitative is the classic left versus right brain operation clash. Instead of picking a side in research, the researcher can use both to get the best of both worlds.

What should be the proper application sequence of the two methods in research? Should the qualitative lead to the quantitative method, or vice versa? Another simple way to ask the question is should "what is" or "how much" lead to the research question? The golden rule of using the mixed method of research is qualitative first and quantitative second for the simple reason that the researcher needs to know "what is" before he can measure "how much."

A supermarket is recently not doing well in business, so they conduct a customer qualitative survey to identify the probable cause to include customer service, products, store location, store hours, price, and so on. At the completion of the survey, the store identifies customer satisfaction as a recurring pattern. The store launches a subsequent quantitative survey to include a series of in-depth questions to get more measurable ratings on specific aspects of that experience such as waiting time, friendliness, knowledge of the product, and so forth. In this example, the golden rule of using the mixed method is demonstrated.

In the previous chapter, we fly over the history of science education in a flash of more than fifty years. In this chapter, we will make two strategic steps to appraise the impact of the historical science education initiatives, one being the qualitative post-Sputnik science education "audit" and the other being the quantitative pre-Next Generation Science Standards (NGSS) "audit." Please understand that the word "audit" is used here to mean checking work in progress.

In figure 2.1, two National Science Foundation (NSF)-funded audits and three student science education assessments are added to the timeline graph

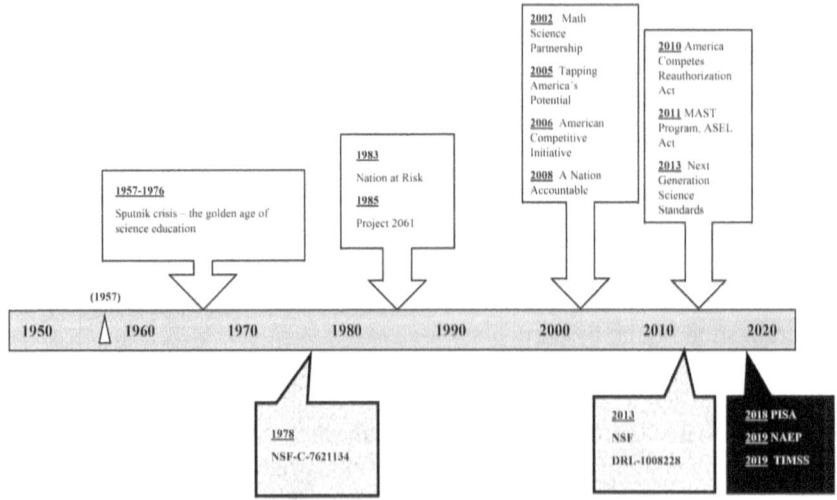

Figure 2.1 **The Science Education Initiative Audits**. *Source*: Author created.

in chapter 1. The two audits are thirty-five years apart. The 1978 NSF project was conducted by a team of researchers from the University of Illinois, Urbana-Champaign. It represents the qualitative method of research. The 2013 NSF project was conducted by the Horizon Research Inc., a leading nationwide research firm specialized in testing and assessment. It represents the quantitative method of research.

Please note that the two projects are meticulously chosen because they are handpicked by the rigorous NSF Request for Proposal committee to promote science, mathematics, and engineering excellence in the United States as a global research and innovation leader. The golden rule of qualitative before quantitative is followed with reference to the order of the two investigations.

Lastly, one national assessment of science education report (i.e., 2019 NAEP) and two international assessment of science education reports (i.e., 2019 PISA and 2019 TIMSS) are added in figure 2.1 to purposefully assess the science education achievement of students.

THE QUALITATIVE INVESTIGATION (NSF-C-7621134)

Today, the education world is more complex ever than before with reference to ethnography and resource distribution. Educators struggle to understand why and how students and teachers feel, think, and behave the way that they do. Many education researchers use the qualitative method to investigate problems and issues. The research information will later lead to important decisions to include but not limited to the effectiveness of current policy or defining new policy to better meet the various constituent needs of the system.

Have you ever wondered why a historical case study is relevant to what we need to know today? The answer is simple that we learn from how people approached issues and found answers to help us move forward. Is this not the wise life's philosophy of "learn from the past to plan for the future"?

Case study research methodology is one important form of the qualitative method used in a number of disciplines including education, social sciences, business, law, health, and many others. It has developed over the years using strategies such as observation, focus group, and interview to study trends and for explaining the operation of a school down to its classrooms based on real-world experiences. This is similar to reporting what we see which is experiential versus measuring what we see which is rationalistic. Case studies tell a story of what we see. Let us follow a tell-tale science education case study going back to 1978.

At this time let us distinguish between a case study and a multiple case study. The former is the persistent study of a single case and the latter is a collection of individual case studies where one tries to make the several case studies alike in some ways to provide a synthesis of findings to understand the whole collection.

A telling 1978 multiple case study (Stake et al.) was conducted to investigate the science education environment in public schools. The investigation made the effort to answer the important questions: "Is science education given a high priority of learning and teaching in public education?" and "Is the national program foundational support aligned with the national needs of science education?"

What is interesting is that the questions are honestly not time-bound in a way that we can also ask public education today. Throughout history it takes a national crisis to get our attention to subsequently kick-start efforts to seek solutions. The 1978 multiple case study was well-timed because the 1970s was the critical post-Sputnik era for the stern global science and math education steeplechase.

The case studies investigation was funded by the prestigious NSF led by a team of researchers at the University of Illinois, Urbana-Champaign. A national stratified-random-sample representing eleven geographical locations across the nation was surveyed. For that reason, the multiple case study consisted of eleven individual case studies. The monumental research took eighteen months actual time and about six research person-years to complete.

The 1978 case studies assessed the status of science education in American public schools. There were three phases of investigation to include: firstly, the on-site observation of case studies of conditions and characteristics of science education in eleven school districts carefully selected by ethnographic characteristics of peoples, cultures, customs, and habits; secondly, site visit to the same eleven districts by project personnel and specialists in science education; and thirdly, a national survey.

The findings of the 506-page report can be highlighted from general to specific information below. They are the general education goals issues, general education issues, and the exclusive science education issues.

(I) Education Goals Issues

1. Total enrollment in public elementary and secondary schools grew rapidly during the 1950s and 1960s, reaching a peak year in 1971. This growth in enrollment reflected what is known as the "baby boom," a dramatic increase in births following World War II.
2. The number of teachers is a function of the number of children there are to teach; the number pattern determined teacher hiring and laying off.
3. Schools are the agents of change and paradoxically they are also the deterrents to change.

4. The purposes education can be placed on serving the human purpose, the knowledge purpose, and the career purpose. The knowledge purpose seems to take the emphasis among the three. However, how to achieve the three purposes equitably draws a controversial debate.
5. Teachers deal with both academic and nonacademic needs of students such as striving for a sense of racial or ethnic identity while retaining a core sense of societal cohesiveness.
6. Teachers, and specially high school teachers, oriented their teaching toward academics to prepare students for the next level.
7. It is hard to make students understand that life requires youngsters to do a lot of learning that do not make sense at the time. How do educators convince students that those learning experiences will ultimately add to the collective wisdom of the person to tackle the many life's twists, turns, and detours?
8. Schools are reflected by the people connected to it. Schools are pulled in many different directions as there are diverse opinions in the various people groups that it represents.
9. Schools want to teach what parents and students believe is useful. Universities want to stress theoretical ideas to search for truth. How do we compromise the school and the university viewpoints? How do universities offer courses that are appropriate to the times to fulfill teachers' need?

(II) General Education Issues

1. The majority of the teachers with special reference to the elementary grades are white females. Will this exclusive workforce of "Teacher Barbie" adequately serve the ever-increasing diverse student population?
2. Many fresh-out-of-college teachers are not adequately prepared to deal with inner-city school responsibilities.
3. The change in student demographics in large metropolitan areas challenges the appropriate application of pedagogy. The American aspiration faces steep encounters in schools.
4. School funding is inequitable. Increased expectation of the schools is seldom matched with proportionate increases in funding support.
5. School accountability by funding source requires teachers to worry about filling out many paper forms in addition to classroom duties.
6. Societal problems are infiltrating schools profusely. Precollege education may be asked to do things once left to the Almighty.
7. Teachers face many youngsters with poor work ethics, little respect for school authority, and meager appetite for self-empowerment through academic learning.

8. The realistic assessment in student learning is probably behind the strong administrator support of the essential "back to the basics" curriculum. Subsequently, many tend to stick to areas where their success has been documented, to the teaching of English language arts and mathematics.
9. Teachers have formed collective organizations as a means of job protection as well as monetary advancement. The truth of the matter is that these organizations protect under their wings both competent and incompetent teachers.
10. Youngsters become increasingly disillusioned by all the glitters and glimmers. Cars continue to serve as the obvious artifact of the youth culture; thus, working while attending school to support the culture is inadvertently acceptable. What is the impact? Some schools are increasingly tolerant of youngsters working full or part time, and even make accommodations for late class arrival and early departure.
11. College-bound high schoolers value the worth of grade point average more than the worth of learning itself.
12. Parents tend to support the importance of college prerequisite courses over elective high school courses.
13. The impact of exclusiveness as seen in the ability grouping of students for instruction is questionable. Is this a proven effective pedagogy for the convenience of the adults at the expense of the students?
14. There is taunting between student tracking and grouping. Grouping is meant to be a temporary assignment to learning groups. Tracking, as in general, advanced and honor is meant to be more permanent as in the different level placement of the same course.
15. The teacher seniority system supports the experienced teacher to teach the bright and able students, leaving the academic less capable students to the inexperienced teachers.
16. What is the latitude of academic freedom before teachers get into trouble?
17. There is a standoff between the boredom of the instructor against the immaturity of the learners.
18. Learning by experience such as field trips as opposed to book learning is considered at the elementary level to be a good change of pace.
19. There is a small use of out-of-school activities such as field trips, camp, museum, and so on to stretch the mind of learning. Could this be a funding or time issue?
20. The growth of technology in our society was important. It represented the quality of life in the country.
21. The use of instructional technology is gaining momentum to assist procedural technologization and prepackaged individualized learning.

22. The source of knowledge authority is the printed information in the textbook more so than the teacher.
23. There is a strong philosophical bias toward the authority of book learning that many teachers at the elementary level especially believe children should be disciplined to learn expeditiously from printed materials.
24. Student management remains a top priority in school issues.
25. There is a mix of opinions on achieving uniformity in schools with reference to curriculum and instruction recognizing that too much of it can be as troublesome as too little.
26. High schools say that junior high schools do not prepare students well, the junior high schools complain that the elementary schools do not prepare the students well enough and the elementary schools complain about the parents. Is this not a traditional blame of vertical articulation? How do schools determine mastery of learning properly to avoid social promotion?
27. Student involvement in schools varied in a blend of motivation, experience, and performance. Student academic motivation was not a simple challenge in learning. Student attitude toward learning was a prime concern.
28. Many teachers commented that test results did not tell them more than what they did not already know about the students. Consequently, teachers made little use of student assessment data to adjust instruction.
29. The diversity of student motivation, experience, and performance affected all aspects of student learning.
30. The education of ethnic minority students was a concern as a group of students influences the performance of teachers and their instructional expectations.
31. There is a disconnect between subject-matter knowledge when transformed in the classroom to meet the socialization demands of the schools. In other words, the academic intellectuality does not connect to the students for proper behavioral responses.
32. Teacher-centered teaching is the easiest form of delivery for everybody. Teachers appeared to use subject matter to demonstrate their competence and social status as a professional educator.
33. People are open-minded when it came to the teaching style and the inclusion of topics dealing with controversial subjects. Teachers have the right to express their opinions although they should also discuss alternative views.
34. Extrinsic motivation is more important than intrinsic motivation if students are to pay attention to their school work.
35. Regular assignments and testings are essential to student learning.

36. Teachers can generally be categorized as being a stern, a liberal, or a mix of both to accomplish socialization goals. The stern socializer promoted subordination, discipline, and competitiveness. The liberal socializer promoted skepticism, imagination, individual expression, cheerfulness, and cooperation. Schools are what people make them, and teachers were socially oriented more than subject matter or student-oriented.
37. Student achievement testing is only one way schools attempt to meet the accountability account.

(III) Science Education Issues

1. Science education is seldom promoted as a matter of sound education. English language arts and mathematics seemed to rank higher with reference to the overall effectiveness of education. Students cannot learn science until they have shown proficiencies in reading and math.
2. When talking about accountability, teachers often reference minimum proficiency in math and reading. The topic of science education almost never came up in the educator conversation.
3. At the elementary level science does not get a substantial share of the school curriculum and the good job of teaching it.
4. The discovery science teaching only works with the academically able students. This is where teachers do not tell students the answer. Approximately, 10 percent of time was spent in inquiry teaching in which students design and carry out their own investigation.
5. There is the perception of a lack of science relevance to various career goals.
6. There is a tug between studying science and getting a job versus studying something else and getting a better-paid job after high school graduation.
7. Educator and parents wanted all students to take some science courses, yet most were not supportive of changing the courses to fit those who are not college bound.
8. The average elementary school teacher was not prepared to teach science. Many teachers teach science the way that they were taught.
9. The delivery of science in the classroom is commonly a teacher show and tell mode of teaching.
10. Elementary teachers were comfortable in teaching science facts and knowledge and less in understanding the relationships among the facts to form the core concept to be learned by students.
11. Tests are public manifestations of understanding. Teachers wanted their students to understand the science subject matter by doing well on the tests.

12. Very little inquiry learning or discovery teaching was observed as a quest for understanding. Science lessons were not aimed at critical thinking, but for discovering what others have discovered to promote what experts have come to accept as standard conceptualization or theory. The back-to-the-basics seemed in many ways directly opposed to student inquiry. Teachers felt that they need more preparation time for inquiry teaching and went down the teaching path that demands less preparation time.
13. Many science teachers felt that the emphasis on facts and techniques has been about right, and there was not enough emphasis on the development of science concepts.
14. Teachers of higher level science courses such as chemistry and physics tend to breed the motivated students in the academics so they can be compared to other schools in the area. These science teachers are there to teach the elite of the school to foster exclusivity. Comparatively, science at the elementary level is taught without the air of prestige and exclusivity. Could this be the effect of heterogenous and homogenous grouping of students for instruction?
15. Is science education value free, and how can science be taught to uphold the American value?
16. There is a lack of interdisciplinary efforts in the instruction of science. In other words, physics is physics, chemistry is chemistry, biology is biology, and little else.
17. Laboratory work is diminishing in importance due to the many management problems of students and the access to supplies and equipment.
18. Many students do not develop a serious interest in science as a vocation.
19. The major causes of high school student dissatisfaction with science courses were students' immaturity, subject matter is irrelevant to student lives, and boring lessons.
20. Science students are trained to find the right answer, and learning is not a matter of experiencing.
21. There is little funding support for prepackaged science instructional material so the teacher's preparation time for hands-on science can be decreased.
22. Many high school science laboratories were reported to be poorly equipped and run down with reference to gas, water, and supplies.
23. Sex role socialization was perceived as a primary factor underlying gender differences in science achievement as girls are less career-oriented than boys.
24. Science workshops in professional development were perceived as not real world. Unless workshop instructors come to the teacher's

classroom, work with the students, use the materials, and show that students respond positively, there is honestly little chance of success.
25. Students said that the things that are most right about the science courses that they have taken are that the courses are interesting, and they stressed the basic facts. On the other hand, students said the other things that are most wrong about the science courses that they have taken were that the courses were boring, and they overemphasized facts and memorization. Obviously the opposing opinions do not belong to the same student group surveyed.

The findings of the 1978 research project come in three major layers each to address the goals of education, general education issues, and science education issues, respectively. Figure 2.2 shows that the exclusive science education issues are like the top layer of an onion, while the inclusive education general issues are at a layer below, and the education goal layer at the core.

Regardless of the specific issues we explore in education, another way of looking at the onion analogy is that there are always the buried layers below as the common denominator, and at the core is the lowest common denominator as expressed in the mathematical academic language. This vividly supports the expression that any real-world issues are seldom one layer deep.

The goals of education manifest the general expectations of the society, and the educational expectations have greatly changed since the time of Thomas Jefferson, the third president of the United States from 1801 to 1809. Jefferson believed public education was to prepare citizens with the basic literacy and mathematical skills that they would need to carry on their lives. Today, one can still see similar education goals focusing on literacy and numeracy as the two pillars supporting the common culture of accumulated knowledge.

Currently, people call the support of literacy and numeracy as the back to the basics movement of education. Some enduring qualities of learning and teaching are timeless. They are taught in schools today, and they will also be taught in schools tomorrow. Is this not the meaning of essentialism and perennialism in the philosophy of education?

The general education issues reflect the changing societal expectations. Going back in time, in the 1800s minimal educational skills were needed for

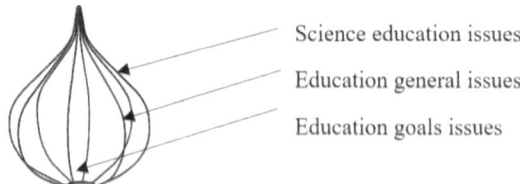

Figure 2.2 **The Layered Issues of Science Education.** *Source*: Author created.

an agrarian society. By the 1900s, there was a dominance of supply and consumption of manufactured goods, requiring an expansion of learning beyond the agrarian society. High school vocational education was a popular option for learning and teaching at the time. Moving into the 2000s, informational technology has become the essential globalization tool.

During 1978, computer-assisted instruction was coming to age. It represents a major tectonic shift in education. Going along with the shift was a swamp of challenges including but not limited to student management, accountability, effective pedagogy, and other invasive societal problems in school culture. Is this not the exhibition of progressivism and social reconstructionism in the philosophy of education?

From the business point of view, the investors are interested to find the input as the process of making an investment, while output is the process of getting something out. An input-output model in business shows the relationship of those input factors (i.e., hefty funding support for personnel, services, and goods) going in leading the system organization to produce the final goods (i.e., student learning performance). The science education issues in this NSF report are noteworthy because the public has the interest to find the learning and teaching output of the national high priority of science education.

In business, the manufacturer rejects the product if it does not meet or exceed the output specifications, a process we call quality control. Can we do the same quality control to students? This is honestly the perplexing million-dollar question.

If money investment is the answer to solve education problems, then why do we still have so many unsolved problems in education? The 1978 project still re-counts science education as a negotiable in the traditional school culture in which English language arts and mathematics remain to be the king and queen of the top education priority.

Teachers, and specially elementary school teachers, are still not well prepared to teach the physical sciences. There is still a fuzzy disconnect between science and the real-world application. In addition, there is a lack of effective pedagogy to motivate teachers to deliver critical thinking type of science investigations. This is indeed a philosophical tug of war between teacher-centered education and student-centered education. If this is not a tug-of-war, then what would be the ideal mix of the two education philosophies to maximize student learning in science?

THE QUANTITATIVE INVESTIGATION
(NSF DRL-1008228)

The 2013 NSF-funded investigation (Trygstad et al.) is the second audit of the science education initiatives handpicked for this chapter shadowing the qualitative first and quantitative second guideline. The study highlights results of the 2012 National Survey of Science and Mathematics Education involving approximately 10,000 K–12 science and mathematics teachers in the fifty states and the District of Columbia. This is the main reason why the survey study with measurable data is by nature quantitative.

The 2013 investigation is conducted thirty-five years after the earlier 1978 investigation to assess whether the nation's schools are ready for the next important debut of the 2013 NGSS. As we know it, the average span of one human generation is about thirty to thirty-five years; therefore, thirty-five years is considered an important time marker.

In the human order of events, one can witness drastic changes in many aspects. For example, the 1994 Motorola StarTac flip phone to the 2022 Galaxy S22 and the 2022 Apple iPhone 12 in the evolution of telecommunication are excellent examples of what the human mind can do in just twenty-eight years. People sometime wonder whether the human mind is capable of achieving similar strides in education and more specifically science education.

In the previous chapter, we learn that NGSS is one major science education initiative that will impact many aspects of science education practices to incorporate curriculum, instruction, and assessment for many years to come. Consequently, the development and implementation of NGSS is a recognized game changer in science education for the simple reason that it is what teachers need to answer the accountability question of what to teach in science every day.

The investigation findings are clustered under the following eight variables in figure 2.3. Amusingly, the overall findings with implications for improvement fit nicely as the top layer of the onion analogy we explained in figure 2.2 because the study is explicitly about the teacher and his/her teaching.

(1) Teacher Demographics

Teachers with special reference to elementary science teachers are primarily middle age white females, and their average science teaching experience is in the range of ten to twenty years. Thirty-five years after the 1978 NSF investigation, the demographics of the science teaching force remains comparable. This piece of information points to the alarming unbalanced demographics between the teacher population and the student population. The teacher demographic data prods the following four important questions.

(1a) If we are to take the state of Illinois as a comparison, the current student population is less than fifty white. How can culturally responsive

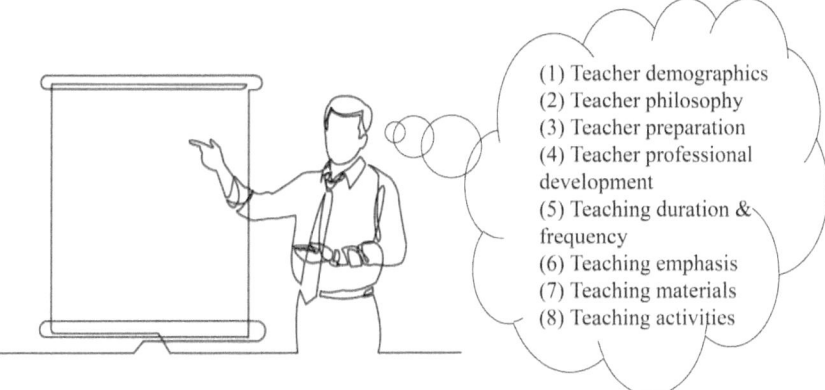

Figure 2.3 The Science Teaching Variables. *Source*: Author created.

teaching be cultivated and encouraged if the teacher demographics and the student demographics are not comparable?

(1b) Can one assume that a teacher with a similar cultural background like his/her students is more effective in teaching?

(1c) Education in general is a white female predominant profession. What does it say about education as a career choice if it does not attract people from different racial groups and genders?

(1d) What does it say about the teacher supply pipeline when the average age of teachers of the study is middle age?

(2) Teaching Philosophy

Teaching philosophy is the intellectual foundation of teachers. However, what teacher believes is one thing, and what they are committed to do can yet be different. In the ideal world, the teaching philosophy reflects both the beliefs and the commitment of doing.

An effective science lesson explained is similar to the hamburger model of teaching. The top bun is a clear statement of the learning purpose and a review of previously learned ideas and skills. The meat patty and the garnish is the learning experience for students to think and reason. Finally, the bottom bun is a summary and assessment of key concepts addressed. The vast majority of the teachers in the study agree with the teaching model described. In addition, they also feel that it is better off to focus on subject-matter ideas in depth other than just superficial broad coverage.

When it comes to the actual lesson delivery, about half of the teachers prefer that they should explain the theory and concept before having students consider claims and give evidence. In that sense, the purpose of the student's hands-on activities is to passively prove the theory or reinforce the concept learned. In a similar teaching manner, the vast majority of the teachers surveyed agree that students should be given the definition of new vocabulary at the start of learning the concept. The data collected in this section of the study prompts the following three questions.

(2a) How does the science lesson delivery in the study align with the research recommendation that students first need to develop understanding of concepts before learning the academic language?
(2b) What is the role of discovery learning and constructivism when the students are not given the role of active learning of the ideas and vocabularies?
(2c) Do teacher beliefs and performance pose a potential barrier to realizing the NGSS?

(3) Teacher Preparation

If science teachers are to be effective, they must themselves have a good grasp of the knowledge as a way of knowing what to teach. With reference to the NGSS framework, the vast majority of the elementary science teachers had taken college coursework in life science to be followed by earth/space science and physical sciences. Sadly, elementary science teachers feel not well prepared to teach all science disciplines.

In a typical college education class, student teachers are more comfortable to teach reading and math, and if they are to pick a science topic that they are comfortable with it will be a life science lesson. This finding is similar to the earlier 1978 investigation. How true is the statement that teachers teach according to how they were taught? This piece of data raises one observation and question and it is:

(3a) Elementary teachers are not adequately prepared to teach all the science disciplines. How do we expect them to follow the NGSS guideline to totally embrace life, earth/space, chemistry, physics, and engineering?

(4) Teacher Professional Development

Professional development is important for teachers not to be behind in what they need to do to perform their teaching duties. It is apparent that a teacher not active in continuous professional development will be behind time in sharing obsolete knowledge and skills.

Let us study the following example to find how teachers can be behind time if they do not keep current in professional development with what they need to know and do in the science classroom. If you took biology in the 1960s, you probably learned about Deoxyribose Nucleic Acid (DNA) as the genetic material and the double helix structure of the DNA molecule.

Moving forward to the late 1990s, the teaching of genetics in biology was more sophisticated to include the genetically modified organism (GMO) breakthrough. The first animal produced using the fascinating cloning technique with a nucleus from an adult donor cell was a sheep named Dolly, born in 1996. Since then the GMO technique is refined and used to improve food production quality and quantity.

How do teachers keep up with the advancement of science knowledge and skills? Continuous professional development is the answer. There are various forms of professional development in which the school can bring the professional development activities to the school, or send teachers outside the school to seek professional development opportunities. Professional development can also be college-level courses, travel, reading/writing books, multiple-grade-level teacher study group, internship with a science education organization, and attending or presenting at professional association conferences.

In the investigation report, over half of the elementary teachers reported that they had participated in science-related professional development in the past three years. In contrast, less than one-sixth of the teachers reported that they had never been a part of any science-related professional development.

Professional development for the elementary teachers is usually more about how to teach science than to deepen the science content knowledge. The report findings recommend that a greater effort is needed to provide teachers with an in-depth knowledge of the NGSS disciplinary core ideas, practices, and cross-cutting concepts.

Do you notice that professional development in science is typically more content-oriented than the teaching of the content. On the other hand, professional development in science education is usually more about the strategies of content teaching than the content itself. For that simple reason, what you learn from attending the American Society for Cell Biology conference is different from what you learn from the National Association of Biology Teacher conference.

When it comes to being an effective science teacher, one needs to know both the content and the content delivery like the person's left and right hand. In the investigation report, locally offered professional development workshops have been topped by state science standards to be followed closely by both science content and science teaching strategies. This finding again reinforced the importance of NGSS referencing the disciplinary core ideas, the practices, and the cross-cutting concepts.

Do you know that textbooks are considered dated if they are older than five years? In a similar token, teachers are also considered dated in their field of teaching if they have not participated in relevant professional development activities in five years. The report findings kindle the following four questions.

(4a) Why is professional development considered an important teacher investment?
(4b) What is an important incentive for teachers to participate in professional development?
(4c) Do you agree with the statement that "learning something new today and not applying it next week is not a good investment in professional development"? Do you have personal example to support your answer?
(4d) How do teachers strike a professional development balance between learning the science content and learning to teach the science content?

(5) Teaching Frequency and Duration

Do you know that the opportunity to learn is connected tightly to the effectiveness of learning? As a professor says at the beginning of the semester that the secret to getting a good course grade is come to class and do the work. Based on a similar logic, one would expect students to learn more science if it is taught more frequently.

The investigation finds that less than a quarter of the elementary teachers teach science on most days every week. This translates to mean that teaching science is few and far between for many teachers. The typical teachers say that they teach science for an average of less than half an hour when they get to it. The teaching time is a sharp contrast to one hour of teaching mathematics and an hour and a half in reading. Moreover, the teaching of science is normally slotted in the afternoon of the day compared to the morning prime time for reading and mathematics. The information here elicits the following five questions.

(5a) How can any science concept be developed and taught properly in thirty minutes?
(5b) Given the inadequate instructional time, does the system encourage teachers to talk more than do science?
(5c) How can more time be given to science instruction? Is this a teacher's decision or an administrative decision?
(5d) How can the system deal with the equitable distribution of instructional time for all core subjects?

(5e) How can one apply the cross-cutting teaching strategy to better make use of the precious instructional time?

(6) Teaching Emphasis

Teachers usually have a concept emphasis in mind when planning for instruction. With this emphasis all the learning experiences are expected to wrap around the concept emphasis as a guidepost of purpose. Borrowing from Professor Ralph Tyler (Tyler, 1949), the father of curriculum development, one would ask "What educational purposes should the teacher seek to attain?" This, by the way, is the first part of the education purpose-experiences-assessment question per Ralph Tyler.

If an atomic structure is set as the lesson concept emphasis, then the subatomic parts of the proton, the neutron, and the electron will be taught to build the structure and support the properties of the atom. In other words, the subatomic particles contribute to building the atom, and the various configurations of the structure contribute to the atom's interactivity with other atoms.

In the study, about half the lower grade elementary teachers agree on the reform-oriented instruction to include understanding science concepts, fostering students' interest in science, and learning science process with real-world applications. Furthermore, the study shows that more than half of the upper grade elementary teachers agree in using the above instructional strategies. This part of the study is encouraging because teachers generally see themselves as working in alignment with the reform-oriented instructional emphasis.

(7) Teaching Materials

Science instruction is inherently driven by supplies and equipment. With limited supplies and equipment, science can only be taught ineffectively by lectures from some printed materials. It is important to realize the close connection between the hands-on and minds-on process as in learning science by doing science.

Put on your science teacher's shoes and walk for a week. How would you coordinate the use of science supplies and equipment in your lesson? To say the least, it will be both expensive and very time consuming. Fortunately, commercially published programs are available to help with the textbooks, equipment, replenishable supplies, and copyable worksheets. The use of commercial published programs translates to one important factor—funding. A good science program is costly because we need to consider beyond the textbooks to include the accompanying supportive instructional resources.

The investigation data reports that close to three-quarters of the science teachers use commercially published programs. In addition, other diligent teachers in the lower grades use other non-commercially published materials as a contrast. All in all, the majority of the teachers are commercially published science program consumers to help guide the overall structure, content emphasis, and pacing of the instructional unit. The information collected at this point raises the following three questions.

(7a) What is the assurance that the textbook is aligned with the NGSS?
(7b) What are the teaching options if the textbook is not aligned with the NGSS?
(7c) What are the pros and cons of teaching out of the textbook and teaching following the science standards?

(8) Teaching Activities

Teaching activities are considered by teachers to be the focus of a typical lesson. When asked what is to be taught in a lesson, many will connect the concept emphasis such as the three states of matter, the water cycle, or seed germination with supportive hands-on activities. In other words, the instructional activities or learning activities to be more student-centered support the intended content emphasis.

Echoing the mind of Ralph Tyler (Tyler, 1949), one would ask "What educational experiences can be provided, and how can they be organized to attain the purpose of instruction?" Here the instructional activities concern the educational experiences and their organization. This is the second component of the education purpose-experiences-assessment question.

The investigation reports that the vast majority of the elementary science teachers use whole-class discussion and explaining science ideas to the whole class as the top two activities. About half of the teachers reported they do hands-on laboratory activities to be followed closely by engaging students in data, charts, and graph analysis. About a little more than one-quarter of the teachers reports that they engage the class in project-based learning activities.

The more severe problem in teaching science with infrequent hands-on activities is compounded by combining the frequency and the duration of science teaching. The combined problem can now be described as approximately less than 10 percent of the elementary teachers teach science using hands-on activities on most days every week. The relatively infrequent instructional activities suggest that the students are given limited exposure to learning by doing scientific investigation as recommended by the NGSS framework.

There are also reports of a small percentage of elementary-level students in self-contained classes receiving science instruction from a science specialist in addition to their regular teacher as enrichment. The findings in this section of the study raise the following two questions:

(8a) Are we adequately preparing our students to compete globally in science education?
(8b) Is there a need to compare our science students with other science students around the world to find where we stand in global science educationachievement?

The 2013 investigation concludes that science teachers with particular reference to elementary science teachers are pathetically not ready for the NGSS initiative due to a major lack of time and resources.

First, what can students honestly learn from an average of twenty minutes per day, three times a week science instruction? Second, how can teachers expect to successfully implement NGSS with limited resources for supporting the instructional activities and the critical professional development opportunities? Third, how can students be active learners when teachers practice passive learning more than encouraging student to be more participatory learners? All in all, there needs to be more teacher support to deepen their science content knowledge and align their practice closer to the NGSS vision.

THE NATIONAL ASSESSMENT OF EDUCATION PROGRESS (NAEP 2019)

Sponging from the cognizance of Ralph Tyler (Tyler, 1949), one would ask "How can we determine whether the purposes are being attained?" to complete the education purpose, experiences, and assessment question. Is this not similar to asking how student learning can be assessed at the end of the lesson? As Michael Piazza in the introduction of the chapter says, audit findings are easy to come up with, but the true value of the audit is truthfully the successful change coming out of it. Subsequently, let us continue with the student achievement change in the wake of the science education audit.

In trigonometry and geometry, triangulation is the process of determining the location of a point by forming triangles to the point from known points. Similarly, in order to construct an accurate big picture of student science achievement in the United States, the process of data triangulation is employed. Figure 2.4 shows the three reference points of the data triangle.

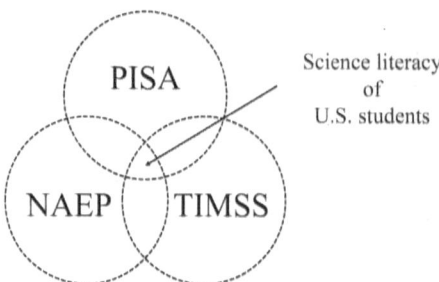

Figure 2.4 The NAEP, PISA, and TIMSS Triangulation. *Source*: Author created.

They are the 2019 National Assessment of Education Progress (NAEP), the 2018 Program for International Student Assessment (PISA), and the 2019 Trends in International Mathematics and Science Study (TIMSS).

Science education assessment at the national level is the systematic broad-scope process of documenting and using empirical data in knowledge, skill, attitudes, and beliefs to refine programs and improve student learning. Such an assessment focuses on the holistic approach of a system. Here, assessment data can be reached directly by examining student's work to assess the achievement of learning purposes. It can also be obtained indirectly from data from inferences about learning.

Please note that assessment carries the meaning of formative versus evaluation which carries the meaning of summative, and the two should not be confused and used interchangeably. For the latest national report on science education, we will first look at the NAEP. NAEP is our first data point of the triangulation process.

NAEP operates under the jurisdiction of the National Center for Education Statistics (NCES). It is a federal body for collecting and analyzing education data in the United States and other nations. A section of NAEP methodically measures the science knowledge and skills of fourth, eighth, and twelfth-grade students with reference to what American students know and can do in three science content areas and four science practices. The three content areas are physical science, life science, and earth and space sciences. The three practices are science principles, scientific inquiry, and technological design. The 2019 NAEP was a combination of digitally based and paper-based assessments.

Table 2.1 shows the overall results of the 2019 NAEP compared to 2015 (https://www.nationsreportcard.gov/highlights/science/2019/). The science scores drop at grade 4, and there are no significant changes in grades 8 and 12 compared to 2015. What is your opinion about the average score of 150 over the total possible of 300? On a widely accepted grading scale of 100 points,

Table 2.1 Average Science NAEP Science Scores 2015 and 2019

Year	Grade 4 average	Grade 8 average	Grade 12 average
2019	151	154	150
2015	154	154	150

Source: https://www.nationsreportcard.gov/highlights/science/2019/

150 over 300 translates to 50 percent. This is a D or more accurately a low D in the report card, and the achievement is nothing to write home about!

It is worth knowing that the 2019 NAEP uses scenario-based tasks for questions. As an example, NAEP has test items connected to life science (4th grade), earth and space sciences (8th grade), and physical science (12th grade). The following grade level tasks are all computer interactive.

In the 4th-grade scenario-based task, test takers are asked to investigate how seeds spread to places where people do not plant them. The students are asked to examine different seed types provided in the task material kit to underscore the skill application of observation, categorization, and making inferences.

In the 8th-grade scenario-based task, test takers are asked to investigate the issue of river water pollution. Here students are asked to perform the water clarity test using the Secchi disc method and later propose ways to alleviate the environmental erosion problem.

In the 12th-grade scenario-based task, test takers are asked to explore using the best materials for a bicycle frame that is strong and lightweight. To do the task, the students need to find the properties of iron, aluminum, and titanium to justify their material selection through the application of the scientific method of investigation.

It is noted that both the NAEP content-type questions and the practice-type questions are well aligned with the NGSS core ideas and practice domains. Can we, at this point, assume the importance of effective science instruction with close connection to the NGSS framework? Well, the 2019 NAEP report notably finds that 4th- and 8th-grade students scored higher whose teachers reported that their students participated more frequently in doing scientific inquiry-related activities. What does this part of the report say about the connection between hands-on and minds-on learning?

The 2019 NAEP finds that less than half of the 12th-grade-level students had taken courses in biology, chemistry, and physics. What does it mean when compared to the graduation requirement big picture of the country?

Science is a graduation requirement for all high schools across the United States intriguingly with disparities. According to the Education Commission of the States (2019), 10 percent of the states require four science courses (i.e., Alabama), 64 percent of the states require three science courses (i.e.,

Texas), 20 percent of the states require two science courses (i.e., California), and 6 percent of the states (i.e., Massachusetts) where the science graduation requirement is locally determined.

If students are to excel in science is it not logical to expect them to take more science courses and for schools to up their science graduation requirement? To further understand the impact of all the science education initiatives of the United States, we need to go beyond NAEP to the next level of international comparison. Let us look at the PISA next.

THE PROGRAM FOR INTERNATIONAL STUDENT ASSESSMENT (PISA 2018)

To better understand the academic strength of our students, we need to do comparison. The PISA is the second data point of the triangulation process described previously for comparison purpose. It is an international assessment that measures fifteen-year-old students' reading, mathematics, and science literacy every three years.

By design, PISA emphasizes functional skills that students have acquired as they get close to finishing compulsory schooling. PISA is coordinated by the Organization for Economic Cooperation and Development (OECD), an intergovernmental organization of industrialized countries and is conducted in the United States by the NCES.

PISA is an international student assessment and for that matter English is used as the assessment language. Do you feel that international students will be disadvantaged if their mother tongue is not English? There is an assumption that the science literacy is based solely on what the students know disregarding their ability to decipher the test questions and communicate their short answer responses in English. For that reason, should the language used in the assessment be factored in the final analysis of student performance?

Science literacy was an assessment area in PISA 2018. The multiple-choice items and short answer are designed to measure students' ability to engage with science-related issues as a scientifically literate citizen. Some sample scenario questions in this area are the pattern of bird migration, meteoroids, and impact craters.

In addition, PISA requires students to know the standard methodological procedures and logic used in science to evaluate or design scientific inquiries and interpret evidence as in tables and graphs. Two sample scenario questions in this area are running in hot weather and bungee jumping. In short, the PISA science assessment is NGSS-aligned in terms of core ideas, practices, and

Table 2.2 PISA 2018 Science Literacy Assessment

Country	SCORE
OECD average	489
China	590
Singapore	551
Macau (China)	544
Estonia	530
Japan	529
USA	502

Source: https://nces.ed.gov/surveys/pisa/pisa2018/pdf/2020166.pdf.

cross-cutting concepts to imply that students well prepared in NGSS should also be well prepared to achieve in PISA!

In 2018, PISA compared seventy-seven countries around the globe. Table 2.2 shows the United States ranked at 18th place in comparison to the top five countries of the PISA report (2019). In other words, the United States was in the top quartile of the report card.

The US average science literacy score of 502 was lower than the average in 11 education systems, higher than the average in 55 education systems, and not measurably different from the average in 11 education systems around the globe. In the United States, 9 percent of fifteen-year-old students in 2018 were top performers in science literacy, scoring at proficiency levels 5 and above; 19 percent were low performers in science literacy, scoring below proficiency level 2.

To better understand the big picture of science literacy assessment, we need to go behind the average test score to find the range of the highest and the lowest score. In the United States, the spread of US student scores in science literacy showed a wide gap of 259 points between the 90th and 10th percentiles. In the ideal world, we like to see a narrow score gap to better reflect the average performance quality of the test taker group. Compared to the earliest comparable PISA score in science in 2006, the average science literacy score of United States in 2018 (502) was higher than the average score in 2006 (489).

After probing the 2018 PISA report, we come down with one conclusion and two questions. The conclusion is that comparatively the United States' fifteen-year-olds achieved better in science literacy despite a wide performance gap between the highest score and the lowest score student. The questions are first what is the chance that the rest of the world school systems are not teaching to the NGSS framework and will that give their students the preparation disadvantage? Second, how can one explain the thought-provoking paradox that the US students performed above average in the 2019 PISA and below average in 2019 NAEP?

TRENDS IN INTERNATIONAL MATHEMATICS AND SCIENCE STUDY (TIMSS 2019)

The TIMSS is the third data point of triangulation as discussed. It specifically provides trend data on science achievement of US students compared to that of students in other countries. For that matter, this is our second comparison study. TIMSS data have been collected from students at grades 4 and 8 every four years since 1995, and the United States has been a participant in every administration of TIMSS. TIMSS is under the administration of the International Association for the Evaluation of Educational Achievement and is conducted in the United States by the NCES.

Based on an analysis of the TIMSS science sample questions (TIMSS, 1995–2011) at grade 4 and grade 8, it is found that the core content of life, physical, and earth and space sciences are covered using the multiple choice and short answer format. Knowledge, application, and reasoning are the three cognitive foundations of the questions.

Many TIMSS questions are what students would read and learn from familiar science textbooks with reference to table, graphs, and diagrams. The test questions support the core science concepts like the properties of matter, earth in space, structure and functions in life organisms, ecosystems, different types of energy, and so on. The general TIMSS test framework is parallel to the NGSS.

How do US students perform in science compared to their international peers (TIMSS, 2021)? In 2019, US 4th-graders' average score on the TIMSS science scale of 539 was higher than the average scores of their peers in 47 education systems and lower than the scores of those in 7 education systems. In 2019, US 8th-graders' average score on the TIMSS science scale of 522 was higher than the average scores of their peers in 26 education systems and lower than the scores of those in 10 education systems. The total TIMSS possible score points are 1,000, and the scale center point is 500.

In the 4th-grade TIMSS test, Moscow city, Singapore, Korea Republic of, Russian Federation, and Japan are the five top-ranked education systems representing city or country. In the 8th-grade TIMSS test, Singapore, Chinese-Tapei, Japan, Moscow city Russia, and Korea republic are the five top-ranked education system. Are you not curious to find out why Moscow as a city, Singapore, Japan, and Korea as countries are top scorers in the TIMSS assessment? What can we possibly learn from them?

How wide is the score gap between the top and bottom performers in science in the United States (TIMSS, 2021)? This is a question that researchers are interested to find because there is always more to be revealed behind the average score.

In the United States, the score gap between top- and bottom-performing 4th graders on the TIMSS science scale in 2019 was 214 points. This score gap is measured by the difference between the scores of students at the 90th and 10th percentiles of the distribution. The score gap between top- and bottom-performing 8th graders on the TIMSS science scale was 254 points in 2019. This gap is measured by the difference between the scores of students at the 90th and 10th percentiles of the distribution.

In the 2019 report, in general, the top-scoring countries have narrower score gaps than the bottom-scoring countries. Do you know that the score gap can be one important indication of equity within an education system? In the ideal world, a narrow score gap represents the steady performance of the students and the rigor of the education system.

WHAT IS THE LESSON LEARNED FROM THE TRIANGULATION OF NAEP, PISA, AND TIMSS?

What can one say about a fish swimming in a garden pond? Before one attempts to answer the question, he needs to understand that a garden pond is comparatively a small ecological system. Due to the size and the resources available, the garden pond is likely to be simple with low biodiversity. Going back to the fish in the pond question, the fish can just be described as many other fish in the pond water. Please note that this description has little to do with whether this is a big or small fish, a good or bad eating fish for the simple reason that there is little comparison in a low diversity pond.

Now, place the same garden pond fish in a freshwater lake, and this lake can be as large as Lake Michigan in the US midwest. A lake is a large body of water, and the resources are plentiful to support a highly competitive biodiversified fish population. What can one say about the fish now? Is it a big or small fish, a good or not so good eating fish?

Do you remember people saying that they rather be a fish in a small pond than a fish in a big ocean? This saying is based on relative comparison. The fish in the garden pond and the same fish in a lake are a comparative analogy that we can apply to student achievement in different education settings. We have heard from our own teaching experience that a top high school performing student can become an average performing student in college, or vice versa.

Similarly, the science performance of the US students in NAEP is in the pond education system in the United States nationally whereas the performance of the US students in PISA and TIMSS is in the lake education system internationally. To recapitulate the results of the 2019 NAEP, the 2018 PISA, and the 2019 TIMSS, US students, modestly speaking, are average nationally and above average in science performance internationally.

CHAPTER TAKEAWAY

What we know about our students in science education over the past many years is a good point of departure for future improvement. We can say that we know where we are now and will use that as a reference point for where we want to go next. We want to go from point A to point B in science education, and point A is now defined and identified.

What our students know and do steer the fate of the country; the next generation represents our hope for peace and prosperity in the future. We need to invest wisely in the minds of our students first and foremost if the United States is to be the world leader. In this chapter we focus on the nation's continuous investment in science education as described in the many national initiatives over time. Tons of resources in terms of dollars are poured into defining new policies and training to support teachers.

Honestly speaking if the resources are the answer to the many challenges of education we would have solved that already in a timely manner. Unfortunately, the investment is going into developing the complex being of the illusive human mind of learning and teaching. That is why oftentimes educators feel like track running in front of a fast-moving train that they have to keep up with.

In the chapter, we describe a landmark NSF investigation report and another strategic NSF quantitative investigation report to delineate the status of science education. The 2019 NAEP, 2019 PISA, and 2019 TIMSS are used respectively to audit the science education status progress. The reading and interpretation of investigation and audit reports are challenging; however, the true challenge toward the end is the successful change of learning and teaching in science for the betterment of the next generation.

REFERENCES

Education Commission of the States—50 state comparison of high school graduation requirements. (2019). Retrieved on May 22, 2022 from https://reports.ecs.org/comparisons/high-school-graduation-requirements-01.

NAEP report card: 2019 NAEP science assessment. (2019). Retrieved on May 15, 2022 from https://www.nationsreportcard.gov/highlights/science/2019/.

PISA 2018: The top rated countries. (2019). Retrieved on May 20, 2022 from https://www.statista.com/chart/7104/pisa-top-rated-countries-regions-2016/.

Stake, Robert E., et al. (1978). Case studies in science education, volume II: Design, overview and general findings. Retrieved on January 13, 2022 from https://files.eric.ed.gov/fulltext/ED166059.pdf.

TIMSS. (2021). TIMSS 2019 U.S. Results. Retrieved on May 22, 2022 from https://nces.ed.gov/timss/results19/index.asp#/math/intlcompare.

TIMSS. (1995–2011). TIMSS Released Questions. Retrieved on May 23, 2022 from https://nces.ed.gov/timss/released-questions.asp.

Trygstad, Peggy J., et al. (2013). The status of elementary science education: are we ready for the next generation science standards? Retrieved on May 10, 2022 from https://files.eric.ed.gov/fulltext/ED548249.pdf?msclkid=d73e472fd14711ecbdaf481e60515164.

Tyler, R. (1949). *Basic Principles of Curriculum and Instruction.* University of Chicago Press.

Chapter 3

Executing the Intended Purpose of a Science Education Policy

Words do not define works.

Maria Karvouni

ANTICIPATORY QUESTIONS

(1) Why do we have office policies in the workplace?
(2) How is the workplace policy developed and executed?
(3) How do we determine the success of a science education policy?

It is mind-boggling to see how we want to see things done compared to how things are actually done because the two might not be the same. In this chapter, we will follow the long and winding journey of policy development to policy execution through several layers of education agency representing the federal, state, and local governments.

Do you recollect a personal experience of giving and receiving orders? Let us see if you remember the following scenarios in a family restaurant.

SCENARIO I

Server: "Good morning! What would you like to have this morning?"
Customer A: "Good morning, I would like to have a combo breakfast with eggs, meat, and toast as shown here in the breakfast menu. By the way, I am in a rush this morning."
Server: "O.K. and do you want to have your coffee now?"
Customer A: "Yes please."

Server: "I will be back shortly with your order. Enjoy our freshly brewed coffee!"

  The server fills the cup and in a short time return with the breakfast platter and a big smile.

Server: "Here is the breakfast. May I refresh your coffee cup?"

Customer A: "Excuse me. Are you sure this is my order? I want my eggs scrambled, ham, and wheat toast with butter on the side."

Server: "I apologize but two eggs, two sausage links, and buttered white toast is shown here in the breakfast menu. Did you not say as shown in the breakfast menu?"

Customer A: "I have no time to argue with you, but this is not what I want. I will make sure that I will not return here ever!"

The customer finished the breakfast hurriedly and left the server with no tip. The floor manager saw the episode and overheard everything. In no time he called the server to the side and gave him advice.

SCENARIO II

Server: "Good morning! What would you like to have this morning for breakfast?"

Customer B: "Good morning to you too! I would like to have a combo breakfast with eggs, meat, and toast."

Server: "Good choice. How would you like to have your eggs? What is your choice of meat? Do you like white or wheat toast?"

Customer B: "I would like to have my eggs scrambled, sausage links, wheat toast with butter on the side please."

Server: "Very good. Let me repeat your order. You ordered scrambled eggs, sausage links, wheat toast with butter on the side. Is this order correct?"

Customer B: "That is correct."

Server: "Would you like to have a cup of coffee or other beverage this morning to go with your breakfast?"

Customer B: "Black coffee would be fine."

Server: "Here we go and the coffee is freshly brewed just this morning. I will be back shortly with your order."

Customer B: "Thank you."

  After a short time, the waiter returns with the breakfast order and refreshes the half empty coffee cup with a smile.

Server: "Enjoy your breakfast and please let me know if you need anything else."

Customer B: "Ok and thank you."

Customer B finished the food and left a 20 percent tip on the table before leaving the seat.

Is this not obvious that customer A in scenario I was not satisfied to put it mildly and customer B in scenario II was satisfied? Let us study the two scenarios more and do a comparison.

There are four players in both scenarios. In the first one, customer A gives the food order, the server receives the order and in the kitchen the short order cook executes the order. Later the floor manager intervened and gave advice. To keep all the players on the same page, the string of communication needs to be accurate and faithfully carried out. It will be wishful thinking if we assume that our everyday giving, receiving, and executing the order can just be a straightforward piece of communication. Unfortunately, it is not! Messages are frequently assumed and misunderstood in the process.

In the second scenario, customer B is a new player. The server, the floor manager, and the cook are the same as in scenario I. Please note that the server carried the customer's order differently following the advice of the floor manager.

What is the manager's little secret advice for the server to change his serving behavior? Notice, in scenario II, the server repeated the order and made sure that it was correct before taking it to the kitchen. The giving and carrying out the order in essence is a communication chain.

In the real work world, communication typically goes through layers of people so the original message is likely to be misinterpreted and filtered as seen in the Chinese whisper game. The game has a line or circle of players whispering a message from one to another to find the communication precision passing on in a fun way. A typical game enjoyment is that slips usually accrue in the retellings, so the message received by the last player typically differs from that of the first player, making it amusing and humorous.

A COMMUNICATION PARADIGM OF ENCODING AND DECODING

Two persons talking or texting to each other is a simple two-way form of communication and any more communicators will add distraction or noise to complicate the process. One form of complex real-life communication is the operational interactions within an organization involving layers of people. The convolution of the process is theoretically proportional to the layers that the information needs to go through, and this includes but is not limited to the setting and execution of an organizational policy.

Figure 3.1 shows a communication paradigm of encoding and decoding. It will be used to describe the setting and execution of the science education policy. The two major players in the paradigm are the encoder sending the message and the decoder receiving the message. The role of the encoder and

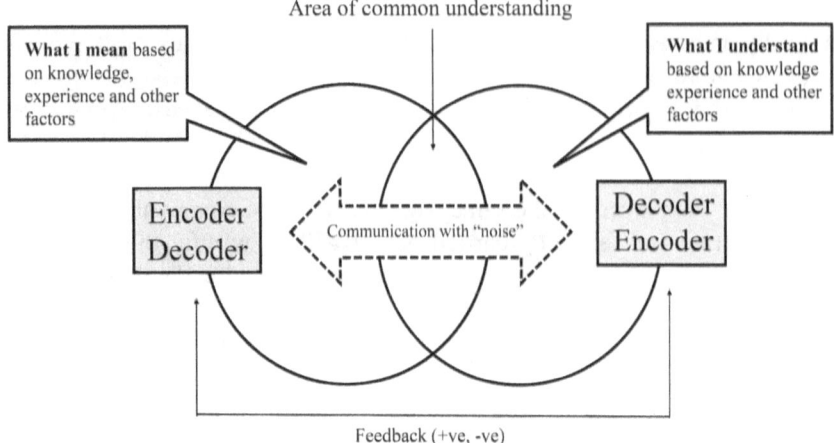

Figure 3.1 A Communication Paradigm of Encoding and Decoding. *Source*: Author created.

the decoder will subsequently be reversed to make the communication two way.

The formation of the initial message from the encoder does not occur in a vacuum. It is a basic connection between stimulus and response as proposed by Professor Edward Thorndike, a renowned American psychologist and professor at Teachers College, Columbia University. The encoder in the first restaurant scenarios described earlier is the server. His greetings and taking the breakfast order are prompted by following the service policy of the workplace. His subsequent change of behavior in the second scenario is motivated by the dissatisfaction of the customer and the advice by the floor manager.

The crux of what has just been described at this time is depicted on the left-hand side of figure 3.1. The knowledge, the experience, and the other factors constitute the message development of the encoder which in this chapter represents the government at the federal level. Let us study the setting and execution of a science education policy to illustrate a more complicated level of encoding and decoding.

There is a system hierarchy in education for communication encoding and decoding that goes back and forth before it is made official into a policy and finally a law if that is the intended final destination. The established pecking order of the education system hierarchy is the US Department of Education (USED) at the federal level, the State Board of Education (SBOE)/Regional Office of Education (ROE) at the state level, the school district, and the school at the local community level.

In figure 3.1, the double-headed arrow denotes the communication through the layers of government education agencies, and the label "noise" speaks for the challenge of all forms of communication infidelity. Let us explain the duties of the different government agencies in order.

THE US DEPARTMENT OF EDUCATION (USED)

The government will not ask you, but a portion of your income will involuntarily go to support public education. Let us be forthright about the education hierarchy with reference to the connection between the resource support and the level of control. The biggest resource contributor, frankly speaking, tends to have the loudest say in communication.

First, the USED operationally comes in a narrow form of categorical aid with constraints. Second, state support comes in either the form of general aid with no constraints or categorical aid with constraints. Third, local support from local property taxes comes in general aid with no constraints or categorical aid with constraints. For that matter, do you know that of the large annual sum of money spent nationwide on education, a substantial portion comes from state, local, or private sources?

Are you surprised to learn that the federal monetary contribution to education is less than 10 percent (Wong & Casing, 2010)? The small resource support through the USED means that the important task of education is primarily on the big shoulder of the state and local governments. Moreover, the tenth amendment of the US constitution expresses the principle of federalism or states' rights clearly stating that the federal government has only limited powers delegated to it by the constitution, and the remaining powers are entrusted to the individual state.

The USED mission is to foster student achievement and prepare future citizens for global competition by promoting educational excellence and equity through two major channels. First, the department assumes the leadership role in the continuous national dialogue on student learning improvement. This, for example, involves activities like heightening the awareness of the challenges facing science education, distributing the latest research on what works in learning and teaching science, and finally rendering help to schools with solutions to the issues. Second, the department follows the excellence and access through the program administration in the full spectrum of education that ranges from preschool education through postdoctoral research. Let us visit the homepage of USED to find out more.

A visit to a recent USED home webpage (n.d.) displays the taskbar tabs as student loan, grants, laws, and data. We will not dig deep into all the details; nevertheless, we need to know that some of the tabs are "seasonal" while the others are where we can rely on finding the pertinent information about what USED is all about.

Visit the laws section of the USED official home webpage, and you will find all the relevant education rules and regulations. Where can you find policies applicable to science education? Please note that there is a special section for you and me to propose and submit a petition for rulemaking. Does this feature not indicate what "for the people and by the people" is about?

Visit the data section and you will find interesting quick facts supporting the success of educational programs and the achievement of our students nationally and internationally supporting what we just learned in the previous chapter. Moreover, you can always find a topic which is not readily found on the webpage. For example, type in "Next Generation Science Standards" in the search box, and it will connect you instantly to everything you need to know about Next Generation Science Standards (NGSS). This is definitely convincing about the power of conducting a web search.

Putting the Sputnik crisis scenario back in the communication paradigm in figure 3.1, the Sputnik launch is the stimulus, and the delivery of a bigger national defense budget and the overhaul of the education system by the United States are the responses. All the major education initiatives described in the first chapter with the exception of the 1957 Sputnik Crisis were managed at the federal level by the USED. Please note that the USED was established in 1980 as a cabinet level agency; consequently, the Sputnik crisis in 1957 was administered by other state departments.

Today, USED serves in the order of 18,200 school districts and over 50 million students attending approximately 98,000 public schools and 32,000 private schools (https://www2.ed.gov/about/overview/fed/role.html). The stimulus and response in this example are nothing like the reflex of the knee jerk. It takes time initially for the stimulus to build up the concern, the subsequent actionable decision to address the concern, and the final evaluation of the policy outcome.

The substance of what has just been described is on the left-hand side of figure 3.1. The knowledge, the experience, and the other factors constitute the message development of the encoder which in this case represents the government at the federal level. We invite you to visit the USED webpage to learn more.

THE STATE BOARD OF EDUCATION (SBOE)

School administrators and teachers seldom get directives from USED at the federal level mainly because they follow the immediate marching orders directly from the SBOE at the state level. The United States has fifty SBOE not including the District of Columbia and Puerto Rico.

The many federal-level decisions made are not implied to be a suggestion. In education, they filter down to the schools as a mandate. The word "filter" is used because the state which is the next level down in the system hierarchy has the power of the tenth amendment to make her own interpretation in answering the federal accountability. SBOE represents the first decoder of the communication paradigm in figure 3.1.

SBOEs consist of elected officials, and they are nonprofit organizations. Individual SBOE represents the state in national- and state-level policymaking, facilitates the exchange of relevant information, and supports the state in advancing excellence and equity in public education for all students. Take note that the concept of excellence and equity in the statement echoes that of the USED. To better understand what SBOE is about, we need to visit the official web page, and we use the Illinois State Board of Education (ISBE) below only as an example.

A recent visit to the ISBE home page (n.d.) shows a wealth of information appealing to a wide spectrum of professional educators. However, of particular relevance to the science educators are standards and instruction, assessment, and research data reporting.

School administrators and teachers will find a lot of useful information under standards and instruction because the section deals with the business of what to teach and how to teach. The Illinois Learning Standards establish expectations for what all students should know and do by discipline and by grade level. The progression of standards complexity through the grade levels is developed to align with the cognitive ability and concept development of students as they advance. Purposefully the well-developed Illinois Learning Standards (ILE) prepare students for the challenges of college and career.

Go through the list of learning standards patiently by discipline, and you will find the NGSS. The NGSS describe the core content, the practices, and the cross-cutting concepts similar to what has been presented in chapter 1 of the book. NGSS found in the ISBE home page asserts the fact that the standards are clear and non-negotiable expectation from the Department of Education as the encoder to the SBOE as the decoder.

Visit the instruction subsection under standards and instruction, and you will find a plethora of teaching resources. ISBE curates the free resources to assist districts, teachers, and parents to support student learning. These resources do not, however, represent an endorsement or recommendation by the state. For that matter, anything under instruction is only suggestions. If an overseeing agency such as ISBE can mandate educators how to teach, then you and I would not be professionals, would we not?

Is it not logical to visit the assessment section after the standards and instruction next? What we learn from Danielson (2007) is that assessment or learning accountability plan should precede the teaching domain. This is similar to saying we need to decide what we need our students to know as an assessment guideline before we prepare the plan of teaching students what they need to know.

Whether you agree or not student achievement reflects teaching effectiveness, and this is a fact of life in the evaluation of teacher performance. It therefore makes good sense for the prepared teacher to know the high-stake student assessment as in the state examination and other college admission examinations such as American College Testing or Scholastic Aptitude Test. A good understanding of the content and the format of the key examinations helps the teacher to align his instruction to the assessment framework, and it often connects back to the student learning standards.

In compliance with federal testing requirements, Illinois administers a science assessment to students enrolled in a public school district in grades 5, 8, and 11. The three grades represent the last grade level at the elementary, middle, and high school, respectively. The assessment is administered in a paper and online format aligned to the Illinois Learning Standards for Science incorporating the NGSS, which were adopted in 2014.

Each state designs its own state examination for science, and in Illinois it is the Illinois Science Test. The Illinois Learning Standards in Science, incorporating the NGSS, replaced the previous science standards that were adopted in 1997. The NGSS are the most comprehensive science standards that Illinois has ever had. They are more rigorous and detailed as they integrate the content of science with the practices of science. Students do not simply select an answer and fill in the bubble of a multiple-choice test.

The Illinois Science Assessment (ISA) pushes students to apply their knowledge when they give answers, thus better preparing them for higher education and a career. All students are asked to demonstrate what they know; they get to see how knowledge is applied to real-life situations. ISA is aligned with the NGSS reinforcing one more time that the state assessment is standard-driven.

Research data reporting is one important section that we need to visit to understand the big picture of education at the state level. The broad state data include fast facts about the number of school districts, schools, graduation rate, and student mobility. You can even dig deeper to find more specific statistical kind of information about the individual school district or school.

Regardless of which SBOE web page you visit, the academic progress-related section will show you the achievement of students in science. Is the overall students' science achievement proficient or not proficient on the state science assessment? How is the state science assessment report stack up to the national science assessment such as NAEP?

THE REGIONAL OFFICE OF EDUCATION (ROE)

How can one single SBOE possibly serve the many schools adequately in the state? As an example, the ISBE is in Springfield, a good 200 miles south of the metropolitan city of Chicago. Do not worry, there are 56 ROE in Illinois serving over 1,000 public schools; ROEs are the long arms of ISBE.

The office of the Regional Superintendent performs the important functions of regulation and service as directed by the School Code. In addition, ROE acts as an education advocate by providing leadership and disseminating information for educators, school districts, and the public. Service components include the dissemination of information on education legislation, legal issues, cooperative management, research, and administration.

Each ROE service component places the ROE as an intermediary agent bringing together people and resources as in the many science education professional development workshops. Many science workshops such as the NGSS before the COVID pandemic were in-person training, and the mode of delivery has been greatly changed to online.

THE SCHOOL BOARD OF EDUCATION

The local School Board of Education is the second to the last intended decoding destination according to figure 3.1. Depending on the size of the school community, there may be one or more schools in the district. In a traditional school district organization, the assistant superintendent of curriculum and instruction is the official contact person to receive the information from the SBOE. Keep in mind that the assistant superintendent is still not at the level of the school and the classroom until we go down to the school principal and the teachers.

Now we really get to understand the complexity of the communication paradigm. Like before, we will visit a local school district webpage to see how the information is finally disseminated to the teachers and the public. Let us visit a typical local school district webpage to do further exploration in the spirit of science investigation.

The home page of one school district under the science education section states that effective learning and teaching will produce scientifically literate students. These students will ask scientific questions and experiment to make claims with evidence support. Scientific literacy enables students to make informed, responsible decisions affecting their lives in a positive way. Here the school district makes it clear that the philosophy of science education is student-centered with a focus on learning by doing.

The webpage further explains that comprehensive inquiry and science standards in life science, physical science, and earth and space science will be used as the official instructional framework. An integral part of this curriculum is to promote an understanding of the interconnections within the sciences and the interactions among science, technology, society, and the environment. Using comprehensive inquiry implies that students actively construct knowledge by observing, questioning, investigating, problem solving, predicting, evaluating, and communicating ideas as they experience the wonderful world of science

One may have guessed from this webpage description that the science curriculum and instruction of the school district follow the NGSS framework although NGSS is not clearly referenced.

THE SCIENCE CLASSROOM

Lo and behold, the classroom is where the student meets squarely with the instructor, and this is the intended destination of all learning and teaching that we expect to see in education. The students with the effective instruction of the classroom teacher are in the far right-hand side of the double-headed arrow in figure 3.1; they are the intended responders of the communication paradigm.

A science teacher needs to answer two very important questions every day before entering the classroom. The questions are: "What do I teach and how do I teach?" An experienced second-grade teacher proudly points to the science unit guide in the webpage (table 3.1) of her school district and said "I use the unit guide as a framework to plan my instruction objectives and develop the student learning experience accordingly."

The energetic teacher further explained that her teaching is divided broadly into three thematic units representing life science, physical science, and earth and space science with its respective foci. When asked how the three areas are related the teacher replied: "There are common concepts underlying what I teach in science, and they are patterns, cause and effect, structure and function, properties of energy, stability and change, and so on. The common concepts are what I underscore to meaningfully intergrade in the disciplines." The brief conversation with the second-grade teacher was remarkable because she reveals that her teaching follows the NGSS guideline although NGSS was not even directly mentioned.

CHAPTER TAKEAWAY

Classroom learning and teaching do not occur in a vacuum. For a science teacher to teach a unit on rocks to be followed by a field trip to a local stone

Table 3.1 Science Themes of the Elementary Grades

Grade	Life science	Physical science	Earth/Life science
Focus Question: How does science exploration help me to understand the world?			
K	grow and change	constructions	collections from nature
1	connect and examine life	magnets	Earth in space
Focus Question: How does science exploration help me to interact with my world?			
2	life cycles	matter and properties	rocks and landscape
3	habitat	motion	water and landscape
4	human body	electricity	atmosphere and weather, climate
5	nature recyclers	heat, light, and sound	forces in nature

Source: Author created.

quarry is carefully guided by the school curriculum. In addition, the teacher may reference the textbook to reinforce the learning experience. It is likely that the school curriculum and the textbook are guiding the cognitive development of the earth science concept using different approaches like the classification of rocks, the properties of rock, the rock cycle, and the connection between the geosphere and the rest of the natural world.

Do you ever wonder where the science curriculum and textbooks nowadays get their ideas from? The answer is NGSS. If it is not for the development of the standards, science learning, teaching, and assessment will be like a moving vehicle without a Global Positioning System. At this point of the book presentation, we need to recognize that NGSS is a revolutionary policy and is a major game changer in science education.

Idea development will remain as a nice-to-know theoretical concept if it is not pushed further to make it actionable. This is where the developed idea is proposed to become a mandate policy with accountability. The apex of the communication link in education is the federal government. The communication is then passed on to the state and local agencies of education. The state and the local agencies have the "flexibility" to make their policy interpretations as long as they are answerable to the assessment policy.

As Maria Karvouni states at the beginning of the chapter "Words do not define work." It is the people. They are the movers and shakers putting ideas and policy into action. The development and execution of NGSS in science education are good examples to show the encoding and decoding from the USED-SBOE-ROE to the local School board of education and finally to the classroom

Do you have the desire to better yourself, your family, your community, and your country with a new policy or policy change? Think for a moment as a concerned citizen about a change you believe would improve some school conditions in the United States. Private citizens like many of us often find ourselves in advocacy positions, particularly if you are a leader in your private or professional life. Do not underestimate what you can do because you just might be the next person to trigger a new policy change in education.

REFERENCES

Danielson, C. (2007). *Enhancing Professional Practice: A Framework for Teaching.* Alexandria, VA: ASCD.

Illinois State Board of Education home page. (n.d.). Retrieved on May 24, 2022 from https://www.isbe.net/.

U.S. Education Department home page. (n.d.). Retrieved on May 25, 2022 from https://www.ed.gov/.

Wong, O., and Casing, D. (2010). *Equalize Student Achievement: Prioritizing Money and Power.* USA: Rowman & Littlefield Education in partnership with Association of School Business Official International.

Chapter 4

The Intellectual and Social-Emotional Foundation of 2+2 = 4 versus 2+2 = 22

Do not train a child to learn by force or harshness; but direct him to it by what amuses his minds, so what you may be better able to discover with accuracy the peculiar bent of the genius of each.

Plato (428 BC–348 BC)

ANTICIPATORY QUESTIONS

1. What are the value, belief, and commitment in life?
2. How do value, belief, and commitment drive your thinking and doing?
3. How do you describe your value, belief, and commitment in science education?
4. How do what you do in the science classroom align with your value, belief, and commitment?
5. How would you describe the general emotional well-being of your students?
6. How would you describe the general emotional well-being of you, the teacher?
7. How would the emotional well-being of the teacher and the student influence the effectiveness of the classroom?

Do you realize that your value, belief, commitment, and mental wellness all have an important impact on how you think and what you do? Your value, belief, and commitment are the indispensable building blocks of a difficult-to-describe entity we call philosophy, the underlying intellectual foundation of being a professional educator. Mental wellness is also abstract, and it concerns how you deal with yourself and other people. In this chapter, we will look into the tight connection between effective science education strategies and their underlying philosophies and the mental wellness state of teachers.

Let us start off by reading a short interesting story of a teacher and her student, student parents, school principal, and the school board over a math quiz question.

On Monday, Tommy came to see his second-grade teacher after school with his miserable math quiz.

Teacher: "Hello, Tommy come on in. Looks like you have trouble with addition, Let's sit down and go over some of the test mistakes. We all learn from mistakes."
Tommy: "Huh!"
Teacher: "What is 2+2? You wrote 22. When we do math addition we do not just put the numbers next to each other as the answer."
Tommy: "That is stupid."
Teacher: "Alright, if I have two pencils in my left hand and I add another two pencils from my right hand how many pencils do I have now?"
Tommy: "22"
Teacher: "No, Tommy it is 4."

Tommy turned around and ran out of the room.

On Tuesday, Tommy's parents came to see the teacher after school.

Teacher: "It is completely normal for students to be frustrated when they are struggling with a school subject."
Mother: "What is this about Tommy getting some answer wrong in a math quiz of yours?"
Teacher: "We had a class quiz and one of the questions was what is 2+2 and Tommy answered 22."
Father: "And?"
Teacher: "And that is not the correct answer."
Mother: "Says who?"
Teacher: "Says Math."
Father: "Are you calling my son stupid? Who are you to say that my son's answer is wrong."
Teacher: "Oh no.... You cannot honestly tell me that you do not know what 2+2 equals?"
Father: "Tommy is a smart kid, he is a free thinker, and you are restricting his ability to think critically."
Teacher: "I do not feel that we can continue because we are getting nowhere."
Father: "I am going to talk to your school principal and file a complaint."

Wednesday, the school principal came to talk to Mrs. Smith, Tommy's teacher.

Principal: "I understand that you had an issue with the parents of one of our students that he had a wrong answer on his math quiz. I need to be informed when things are getting out of hand because I am the school administrator in charge here."

Teacher: "So how do you want me to resolve the issue?"

Principal: "To apologize . . . it is not our job to tell students when they are right or wrong. Parents do not want your biased view ramped down students' throats to hurt their feeling."

Teacher: "That's exactly what I and other teachers do to tell students how academic works."

On Thursday, Mrs. Smith is summoned to the evening school board meeting.

School superintendent: "Parents are suing about emotional distress to a minor. Please explain your side of the story leading to this fiasco."

Teacher: "I just told a student that 2+2 equals 4."

School board member: "We need you to be open-minded about the possibility that there might be other correct answers."

Teacher: "As a professional teacher I need to hold my ground and I will not be bullied. There is only one correct answer to the math problem. There is nothing to compromise and this is so stupid."

School board president: "For your sake, I have a correct answer for you. The school minus you equals tomorrow! We are suspending you while you are reconsidering your extremist view. Hope you understand that you are doing this to yourself."

On Friday, Mrs. Smith received a morning call from the principal saying that her service is no longer required. She was told that Friday would be her last day of work. She was asked to come to the school on Saturday to pack her school belongings and complete the exit interview.

On Saturday, Mrs. Smith comes to the school principal's office.

Principal: "Thank you so much for coming. Sorry that things happened that way only if you'll be more open-minded."

Teacher: "I have my professional integrity and I am not going to compromise."

Principal: "Let us fulfill our financial obligation here. You have $2,000 for the last pay period and another $2,000 for the next pay period and a total of $4,000."

Teacher: "Wrong!! It is $22,000. $2,000+$2,000 equals $22,000."

Right off the bat, the story above shows a distinct interplay between two opposing views. The first view is teacher-centered, and it is represented by the 2+2=4 math equation. The second one is student-centered, and it is represented by the 2+2=22 math equation. Do you agree that the teacher's mind set and behavior are affected by her professional belief that 2+2=4? On the other hand, do you also not agree that the thinking and behavior of Tommy's parents, the school principal, and the school board are all affected by the alternative belief that 2+2=22?

It is interesting to know that the alternative belief of 2+2=22 though academically incorrect is used figuratively to encourage student's constructive learning; in addition, in our present-day education landscape it is something people used just to be politically correct!

We will not dive into discussing the exclusiveness of the two philosophies at this juncture simply because there are situations in that we need to apply one philosophy exclusively and in other situations we need both philosophies inclusively to be effective. As professional teachers, we need to protect the integrity of the discipline (i.e., 2+2=4) and also find time to tickle the misconception of students (i.e., 2+2=22) and hopefully steer them slowly but surely back to finding the right one.

The philosophy of education is critical in answering two key questions concerning knowledge acquisition and learning effectiveness. Is learning directly or indirectly from experience? Is learning connected to the physical world or is it the internalization of some mental construct? Is this not real world that teachers must take firm stances on answering the questions before they can even determine what they should teach and how students should learn?

In teacher-centered education, students put all of their focus on learning from the teacher. The teacher talks and the students listen. During activities, students' ideas of learning are generally not encouraged. When a classroom operates with student-centered instruction, students and instructors share the focus. Instead of listening to the teacher exclusively, students and teachers interact freely. Students are encouraged to share their ideas in learning.

Do you expect educators to be on the same page regarding the role of learning and teaching? Since educators do not always agree, different philosophies of education have emerged. Now we are ready to explore the finer philosophy divisions beyond being either teacher-centered or student-centered.

Please be assured that your education philosophy matters. You may think more about whether you enjoy being a teacher for the time being, or more specifically whether you are good at being a science teacher. Believe it or not, asking those questions comes from the philosophy that you develop during your professional career. Let us explore five scenarios below to identify your possible science education philosophy affiliation.

ARE YOU A TEACHER-CENTERED ESSENTIALIST?

William Bagley (1874–1946) was a professor at the University of Illinois and later at Columbia University who popularized the concept of essentialism (William Bagley biography, n.d.). He advocated the worth of knowledge for its own sake and underscored the systematic study of traditional academic subjects.

Some people argue that the historical 1957 Sputnik crisis, the 1983 A Nation at Risk report, and the 2001 No Child Left Behind mandate have kept essentialism as the rallying focus of education today. Others agree that the increase in immigration to the United States and the subsequent multiculturalism are challenges to the traditional American identity; therefore, essentialism is important to build and fortify a common culture for the unity of the nation.

Do you know that future teachers, especially elementary teachers, spend the majority of their coursework focused on methods of delivery rather than on mastery of the subject to be taught? A recent American Enterprise Institute study (Arnn, 2022) found that only about one in eight principals and a mere one in fourteen superintendents express confidence in teacher certification. That's alarming! As well-intentioned as many colleges may be, teachers training programs often steer educators away from the subject matter and toward a political agenda. Is this not a concern about teacher candidates not preparing themselves well as essentialists in their traditional academic subjects before entering the teaching profession?

If you are an essentialist, you tend to teach students the essentials of academic knowledge and skills development. This traditional approach to teaching is meant to work the mind for proper reasoning in the application of the content knowledge. An essentialist seeks to lay a common core culture foundation that every student should learn. Many of us have the education background of essentialist schools requiring you to take English, math, science, social studies, and other elective courses. Such school programs are typical essentialism.

Do you lean toward teaching students the accumulated knowledge through core science courses such as chemistry, physics, or biology in the traditional academic sense? What are your preferred ways of instruction? Do you normally introduce a lesson framework to be followed by the lesson presentation, guided practices, and independent practice? This is mostly the acquisition of knowledge through direct instruction, a teaching strategy that goes generally well with a teacher essentialist.

ARE YOU A TEACHER-CENTERED PERENNIALIST?

Mortimer Adler (1902–2001), a professor at Columbia University and the University of Chicago, was a renowned philosopher and a strong advocate of perennialism in education (Adler Mortimer biography, n.d.). The word perennial is descriptive of something that is lasting like the perennial tulip that returns to blooming every spring after the cold harsh winter. In a similar manner, an education perennialist focuses on studying themes or concepts that endure the passage of time.

The perennialist teacher plays great emphasis on reading great books with the attention given to teaching values and character training with additional discussion about the underlying values and principles. Some great book examples from the fictional shelf are *Jane Eyre* by Charlotte Bronte, *Wuthering Heights* by Emily Brontë, *Moby Dick* by Herman Melville, *The Great Gatsby* by F. Scott Fitzgerald, and *Frankenstein* by Mary Shelley. For those who are more science and nonfiction inclined can read *On the Origin of Species* by Charles Darwin, *The Immortal Life of Henretta Lacks* by Rebecca Skloot, and *Pale Blue Dot* and *Cosmos* by Carl Sagan. As Mortimer Adler might suggest, training the mind through reading and discussion is ageless.

As a science teacher, can you incorporate grade-level appropriate great books as part of the science curriculum? How do you select appropriate great book titles and align them to support your instruction? Finally, how do you support science learning through reading and discussing great book ideas?

ARE YOU A STUDENT-CENTERED PROGRESSIVIST?

John Dewey (1859–1952) of the University of Chicago has been recognized as the major scholar reformer associated with wrapping school learning around the real-world experiences of students (John Dewey biography, n.d.). Progressivism is the application of an educational philosophy called pragmatism. A pragmatist determines the merit of an idea through testing and life application. If the idea works in the real world, then it has merit. Do we not often say something that is practical as being pragmatic in English? In progressivism book learning is important, but it is not as important as learning by doing.

Do you know today in America, most of the education is perceived as a cold, life-draining maze of bureaucracy brought over in the early twentieth century from Europe and activated during the wave of progressivism that swept over the United States? Education bureaucracies control more than half the budgets in every state, and they control the education departments where teachers are trained.

Visit the University of Chicago Laboratory School, and you will be rewarded to find for yourself progressivism in action. John Dewey started the Laboratory School as a testing ground for his ideas of experiential learning. The delivery of instruction can be seen in various group methods such as cooperative learning, field trips, role-playing, and dramatization as in social interaction. Do you know that Malia and Sasha Obama attended the University of Chicago Laboratory Schools before Barack Obama moved to Washington, DC, as the president-elect in 2009?

In science education, learning by doing is commonly connected to the scientific method of investigation. The method includes the initial stage of awareness and definition of a problem to be followed by setting hypotheses, collection, and analysis of information as in experimentation to propose the most likely solution to address the problem. Science teachers are familiar with the scientific method, and we will extend the discussion of the strategy in a later chapter.

Are you a progressivist teacher? Invite a colleague to visit your classroom. Does the visitor see you lecturing in front of the classroom most of the time? Or does the visitor see students working in groups, moving around, and talking freely but purposefully about learning? In a progressivist classroom, students might not sit neatly in rows, and what they do is purposeful and not chaotic. In addition, computer simulation and field trips offer real-time learning reinforcement for students and build on their curiosity, ability, and interest. Progressivism taking many teaching elements into account is about student-centered learning and as John Dewey would agree progressivist education is less about preparation for life but more about life itself.

ARE YOU A STUDENT-CENTERED SOCIAL RECONSTRUCTIONIST?

Education has the fundamental roles to transmit culture. When American culture is in a state of crisis, the role of education to modify culture steps up. In view of all the social turmoil in his time, Theodore Brameld (1904–1987) committed his career efforts to employ schools as agents for social re-experience should be reinforced. As an American educator and educational philosopher, Brameld was best known as the founder of social reconstructionism in education (Theodore Brameld biography, n.d.). Social reconstructionism is an education philosophy to encourage the school efforts to alleviate social inequalities and other challenges. In a way, strategies of social reconstructionism extend from meeting the exclusive needs of the learners to the inclusive needs of the society.

Do you know that the University of Michigan—ranked third in the nation for elementary education and eighth for best overall education—declares that its program has a strong emphasis on developing teachers' instructional practices for the purpose of disrupting inequities in schools? Students who earn a degree in education learn teaching practices that adopt a subject-matter serious perspective and are rooted in social justice.

What are some of the social issues that our society is facing these days? There are quite a few current social issues that should be addressed in America; however, the following ten keep bubbling to the top whenever the conversation comes up. On the list and not in any particular order are the Covid-19 pandemic, health care, education, unemployment, environment and climate change, obesity, foreign relations, illegal immigration, sex inequality, and racism.

What do you expect to see when you visit a social reconstructionism classroom and most probably it is a social studies classroom with a social reconstructionist teacher? You would expect to see a teacher modeling the democratic principles by engaging the class in a democratic culture.

The teacher might conduct a provocative lesson that stirs the student about the inequalities that surround them from reading a book or what is current in the news headline. The teacher leads the class discussion, but discussion unaccompanied by action is not what reconstructionism is about. The student-centered reconstructionist teacher assigns students to survey the community perception about a racial relation, or analyze the public school report card to determine the challenges of the local school system, and the assignment list continues. In view of the sample assignments described, the teacher would be coaching students on research techniques, statistical assessment, and writing and reporting skills all in the effort to bring awareness and effort to improve the society.

Is it possible to have social reconstructionism in a science classroom one might ask? The answer is absolutely possible because the nature of a social reconstructionism lesson is determined both by the concept of the lesson and how the lesson is delivered. A social reconstructionist lesson is not the off-limit education territory of a social studies classroom. Let us visit a science classroom next.

In view of the ongoing impact of the Covid-19 pandemic, one concerned science teacher asked his students to collect and analyze the public Covid-19 vaccine record to assess the effectiveness of the local health-care system. The survey found that Covid-19 vaccination in the school community had given low priority to the needs of school children. Is this not a public health issue? It is. Using the survey data, the students under the supervision of the science teacher organized a Friday morning school rally to motivate all students to get their vaccination. Is this not a beautiful social reconstructionism learning experience example to show that when actions are based on data and value of the common good, even the next generation students can make a difference to better the world?

ARE YOU A STUDENT-CENTERED EXISTENTIALIST?

Existentialism is a different philosophy from the four described previously and is the last one that we will discuss in light of education. This philosophy advocates that humans at all time deal with human existence and human essence. In a way, human existence is about life and human essence. It is about all the essential elements of being a human. The challenge of existentialism is the belief that human existence comes before human essence to imply that human existence comes with absolutely no endowment. For that reason, it is up to the individual to find meaning and purpose in life.

Søren Kierkegaard (1813–1855) strongly proposed that there is no destiny in life and is up to the individual to be responsible for making choices for self and other. Søren Kierkegaard is the father of existentialism (Søren Kierkegaard biography, n.d.). He said that the substance of existentialism is individual existence, freedom, and choice. How does existentialism influence the way we learn and the way we teach? Let us visit an existentialist classroom.

In an existentialist classroom, you will find that it is anything but traditional because learning is student-directed and self-paced under the guidance of the coach, the teacher. In that sense, it is the flip side of the essentialist coin. The role of the teacher is to help students to explore their own learning essence by exposing them to various paths that they may take to learn and solve problems. Students are to navigate the uncharted academic territory in which they can freely choose. Any subject matter that might help the student to explore his potential rather than to imitate the established model of learning is encouraged.

Are you a teacher in search of your own freedom in life? Do you creatively choose the way you learn and solve problem? If you are then you can inspire a student to search for his own because it takes one to teach one. If you live in the United States, you do not have to look too far but visit the Sudbury Valley School just outside Boston, Massachusetts, to find what an existentialist school is like. Lest we forget, we need to realize that in the real world we just cannot presuppose that everyone learn the same way, and this can just be a different education philosophy to help some of our students to reach their potential.

DECISIONS, DECISIONS . . .

As an educator, do you see your value, belief, and commitment being close to the philosophies of essentialism, perennialism, progressivism, social reconstructionism, or existentialism that we just described? It might be easy for you to identify your personal education philosophy if you already have a wealth of work experience. On the other hand, if you are a novice educator it might take time to shape and form your personal philosophy understanding that might change a number of times in the course of your professional career.

Can anyone right out of the gate decide whether he is more teacher-centered or more student-centered in your approach to learning and teaching? Consider the following ifs.

If you are teacher-centered you as the teacher is responsible for teaching what is worth knowing. In the selection of reading and discussion resources, you know what to share based on past established valued thinking. You teach a rigorous curriculum to prepare your students to compete with the rest of the world. Similarly, you believe that healthy student academic competition is key to school performance, driving the country's productive workforce.

If you are student-centered, you as the teacher understand that you cannot possibly teach everything so you equip students with research skills and motivate them to learn on their own. Give students the choice to learn best what they desire to learn. Create an environment and let students learn by doing and experiencing the real world. Again, you understand and put in action that meaningful learning reward is best accomplished intrinsically and not by external grades and social approval.

After you read through the above scenarios, are you still not determined about your education philosophy? Let us do the following twenty-item inventory of education philosophies to help. Respond to each of the statements below using a 1 to 5 scale where 1 is disagree strongly, 2 is disagree, 3 is not sure, 4 is agree, and 5 is agree strongly. You will get to interpret the results at the end.

_____ (1) A school syllabus should include a body of common knowledge and skills that all students should know and do.

_____ (2) A school syllabus should concentrate on perpetual timeless great thinking and ideas.

_____ (3) The learning experience should be reinforced by real-world experiences.

_____ (4) Schools need to prepare students to tackle social problems they face beyond the four walls of the classroom.

_____ (5) Give students the choice to determine what they need to learn under the guidance of the teacher.

_____ (6) Students will not be promoted to the next level of learning until they master the key knowledge and skills.

_____ (7) The purpose of schooling should focus on the development of students' creative thinking and problem solving over the learning of current trends and social skills.

_____ (8) Learning needs to connect the relevance of the real world to the classroom.

_____ (9) Learning needs to include the recognition of societal injustices and students responsible to address social inequalities as future citizens.

_____ (10) Schools should offer students the choice of learning something in a way that they are comfortable.

_____ (11) The classroom needs to encourage, respect for authority, discipline, and academic rigor.
_____ (12) The best way to learn is to do it with enjoyment through reading, doing projects, and freely discussing ideas.
_____ (13) The teacher is in charge of determining what is worthy with reference to students' learning needs and experience.
_____ (14) School learning should promote harmonious relationships among different racial and ethnic groups.
_____ (15) The main purpose of schooling is to help students find out their life's purpose and role in the real world.
_____ (16) Schools need to prepare our students academically to compete with the global economy.
_____ (17) Teachers need to teach from great writings, the classics, because important learning and understanding of today's challenges are found in these great books.
_____ (18) Adequate social interaction is key to effective learning in schools.
_____ (19) Students need to know and do something about improving the quality of life for all people.
_____ (20) Give students the options and choices about following different paths of learning to enrich their lives.

Let us add up the responses you did on the survey in table 4.1. Each column of the table is labeled with the philosophy we already discussed. The highest possible score is 20 and the lowest possible is 5. Total scores of 15 indicate strong agreement, and scores below 10 indicate weak agreement with reference to a particular philosophy.

What is your total score in essentialism, perennialism, progressivism, social reconstructionism, and existentialism? Compare the five philosophies. Which one scores the highest/lowest? Rank your philosophy scores in descending order. Does it align with your life's value and belief? When you look at the top two philosophies by score, do you find that as something that you are committed to? Is it possible that your philosophy may be inclusive of the other philosophies that you have a blend of hybrid philosophy? Yes, it is possible.

Table 4.1 The Educational Philosophy Score Table

Essentialism	Perennialism	Progressivism	Social Reconstructionism	Existentialism
1.	2.	3.	4.	5.
6.	7.	8.	9.	10.
11.	12.	13.	14.	15.
16.	17.	18.	19.	20.
Total =	Total =	Total =	Total =	Total =

Source: Author created.

It is reasonable that a teacher might not be able to answer the philosophical teacher-centered or student-centered educator question succinctly because the effective delivery of a lesson plan depends greatly on the maturity and experience of the students. When the students are less mature and inexperienced, the teacher can give them more help, and the lesson delivery will become teacher-centered. On the other hand, when the students are mature and experienced, then the teacher can deliver more independent type of learning, and this is student-centered.

To make the discussion more interesting, a teacher can exercise a mix of two or more philosophies to meet the learning needs of his students. If the person is a mix of the two philosophies then what would be considered a good mix? The short answer is the right mix should achieve the most in student learning, and the teacher should not follow a one size fits all type of philosophy template. In education, it is important to know that teacher-centered and student-centered practice are not exclusive of each other. Rather, they form a continuum.

Figure 4.1 depicts the nature of the philosophy continuum. Inside the upper shaded triangle are the student conditions of the teacher-centered strategies. The conditions are students with limited prior knowledge, learning as passive dependent knowledge and skills acquisition. As one moves from the left-hand side (i.e., broad base of the shaded triangle) to the right-hand side (i.e., the apex of the shaded triangle), the conditions of student dependence figuratively diminish.

The lower unshaded triangle is next. Inside the lower triangle are the student conditions of the student-centered strategies. The conditions are students coming with extensive prior knowledge, learning is an extension and application of the prior knowledge that goes beyond basic competency. As one

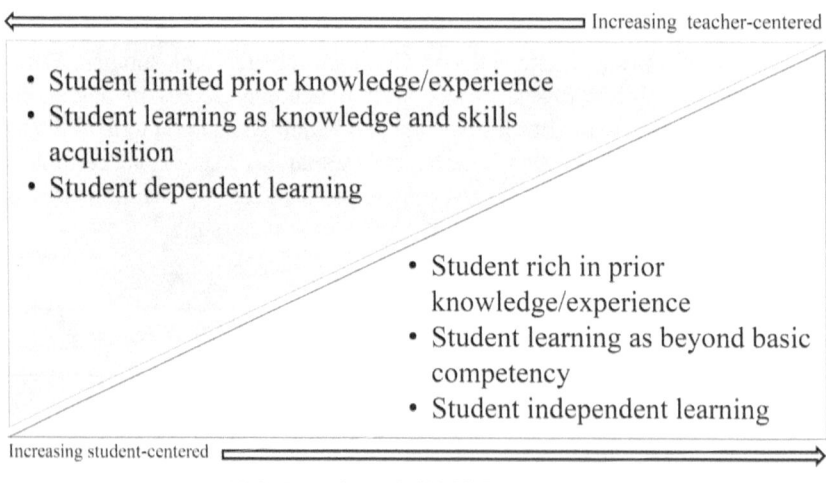

Figure 4.1 The Educational Philosophy Continuum. *Source*: Author created.

moves from the right-hand side (i.e., broad base of the unshaded triangle) to the left-hand side (i.e., the apex of the unshaded triangle) of the triangle, the conditions of student independence figuratively diminish. The arrow at the bottom of figure 4.1 represents the maturity and experience progression of learning through age or time.

In sum, the shaded and unshaded triangles of figure 4.1 are the student condition extremes of teacher-centered and student-centered practices. In the real classroom, the student conditions in a classroom can fall reasonably anywhere between the two extremes. As an illustration, if a group of students in a class shows an even mix of dependent and independent learning characteristics, then the profile of student learning can be placed theoretically right in the middle of the teacher-centered and the student-centered continuum.

Do you truthfully feel that your intellectual foundation matters? You may ask whether you enjoy being a teacher, or more specifically whether you are good at being a science teacher. Believe it or not, triggering those questions comes from knowing and exploring your intellectual foundation.

Which education philosophy is important to support the learning and teaching of the Next Generation Science Standards (NGSS) knowing that they are the science education game changer? Many successful science teachers will share that being an essentialist and a progressivist align well with the expectation of NGSS. If you agree or disagree, will you be able to support that with your own classroom experience?

THE SOCIAL-EMOTIONAL FOUNDATION OF 2+2=4 VERSUS 2+2=22

Let us revisit the story of 2+2=4 and 2+2=22 at the beginning of the chapter. Do you recall that 2+2=4 represents the view of the teacher while 2+2=22 represents the view of the student? What may be the cause behind the 2+2=22 statement could be the student's mathematical misconception, or the student's absurd answer just to get his teacher's attention, or other mental wellness issues.

We will discuss the relevance of misconception in another chapter but will focus on exploring the connection between social-emotional mental wellness and learning. Let us read the next story below to help us kick-start the exploration.

Teacher: "Three youngsters went hiking in the woods and got lost on their way back to camp after sundown. This is a problem that the young boys are up against. How can you help them to get back to camp safely?"

Student A: "Have the boys stay put for the night and wait for the morning to find their way back to camp."

Teacher: "Sounds like a suggestion. Are there any other ideas?"

Student B: "Ok I agree."

Student C: "Do they have a torch light in the backpack? Can they use the light to help them find their way back to the camp? The challenge for them is to find their way home in the dark of the night!"

Teacher: "Looks like another fine idea, and how do we find out how that will work? Any more ideas?"

Student D: "Ok"

Student E: "Ask the boys to go through their backpacks to find what they can use to help."

Student F: "Ok"

Teacher: "One boy found a light bulb, a wire, and a battery in the backpack. Do you think that is helpful?"

Student E: "The three items are not exactly a flashlight, and I am curious if they are really useful."

Student F: "Ok"

Student C: "What if the hikers can connect the battery, the wire, and the bulb to make light? Will that be an idea?"

Teacher: "I have 3 sets of battery, wire, and light bulb. Let us explore the possibility of connecting the items to make light and will do this in three groups of two students. Your little electricity experiment can help the three lost hikers get back to camp."

In this science class scenario, there are enough batteries, wires, and light bulbs for three student groups. How do you help the teacher to group the students, and what is your rationale for making the groups? Based on the class conversation, students A, C, and E are willing and able to share problem-solving ideas. Students B, D, and F on the other hand are passive learners and their passive response is ok and little else. If we are to place students B, D, and E together, we can expect the group to go slow or stagnant with the battery, wire, and bulb experiment.

One grouping suggestion is to separate students B, D, and F and have them pair with student A, C, or E, respectively. Another student grouping option is A and B, C and D, E and F. Other options can be considered as long as B, D, and F are not in the same group.

We will not know the real reasons behind the reluctance of the three students B, D, and F. We need to be reminded that reluctant class participation or other unusual behavior may not be a concern if it is not persistent over time. Assuming that the students do not have special learning needs, the reasons for unusual learning behavior could be a combination of having no interest or being socially shy, a lack of prior knowledge and experience, English language learners, or some other social-emotional behaviors.

Problematic social and emotional behavior is the process through which children and adults alike have problems with understanding and managing emotions, set and achieve positive goals, feel and show empathy for others, establish and maintain positive relationships, and make responsible decisions.

Student's social emotionality is very important in education today ever than before. The mental wellness is about student confidence and engagement that translates eventually to overall school success. Let us visit Hoover High School, Sequoia Middle School, and Jackson Elementary School in Fresno, California, to see how the social-emotional challenge of students that we have traditionally seen as nonacademic had critically affected student achievement (Toch & Miller, 2019).

In a nutshell, the study report showed that 33 percent of students believed that they could master the toughest topics in their classes. In addition, less than 50 percent of students responded positively to the question, "I feel like I am part of this school." It is apparent that there is a lot more going on behind the achievement of students. Is there a possible connection between the pathetic mental state of students and the high suspension and expulsion rates in addition to just 20 percent of its students mastering the California's math standards? Let us find out more.

In 2017, a time before the global COVID-19 pandemic, the Fresno consolidated school district in California with the assistance of external resources conducted an SEL survey of students by grade. The grades 4 through 12 survey conducted a comprehensive questionnaire referencing student growth mindset, student self-efficacy, student self-management, and student social awareness.

In the growth mindset area, there were questions about whether the student has learning obstacles and whether taking on challenges will or will not make the student smarter. In self-efficacy, there were questions about the student's confidence to meet the learning goals to earn a top grade in the class.

In self-management, questions were asked about whether the student came to class prepared, or allowed other students to speak without interruption.

In social awareness, questions were asked whether the student was able to stand up for himself without putting other students down, or listen to other people's point of view carefully. In sum, the social-emotional elements address the motivation of student to engage in school work and activities.

Table 4.2 shows the percentage of positive responses of the survey peaks at the middle elementary grade and drops to the lowest at the beginning of high

Table 4.2 Percentage of Positive Responses to Social-Emotional Questionnaires by Grade Level

	% of positive responses to social-emotional questionnaires by grade								
	Grade 4	Grade 5	Grade 6	Grade 7	Grade 8	Grade 9	Grade 10	Grade 11	Grade 12
Self-manage	58	57	55	46	44	38	36	36	39
Growth mindset	64	69	67	62	64	60	63	64	65
Self-efficacy	60	68	56	46	44	37	36	38	45
Social awareness	72	70	66	58	54	52	56	58	60

Source: Author created.

school to mean that students became less confident in themselves as learners progress through the grade levels.

Interestingly, this downward spiral pattern of student self-efficacy is not unique to Fresno. The survey data are further complicated by factors such as gender, race, and the socioeconomic status of the respondents. These results could reflect larger social influences affecting how children of color and those living in poverty view themselves. Overall, the big picture view of the survey shows that students had a grave problem in maintaining healthy social-emotional stability. How do we honestly expect student to learn and achieve in the reported debilitating mental state?

In March 2020, the World Health Organization (WHO) declared COVID-19 a pandemic. In July, officials debated the best scenarios to allow students to safely return to school in the fall. In August, COVID-19 was reported to be the third leading cause of death in the United States (after heart disease and cancer). In September, the school year opened with a mixed plan to keep children and teachers safe, ranging from in-person classes to remote schooling to hybrid models. With the outbreak of the global pandemic and its significant school disruption, people anticipated the worse for the mental wellness of students.

In 2020, the National Alliance on Mental Illness (NAMI) issued a report (NAMI, 2020) regarding the mental wellness of youth and young adults in the United States going through one of a kind COVID-19 challenge of the disruption of many life routines to include isolation from friends and learning online.

The report recognized the significant negative impact on mental wellness. Among the adolescents (aged twelve to seventeen) 17 percent experienced major depression, and there was a 31 percent increase in mental-health-related emergency department visits. Among the young adults (aged eighteen to twenty-five) 33 percent experienced some form of mental illness with 10 percent being serious. Overall, 20 percent of young people reported that the pandemic had a significant negative impact on their mental well-being. To cope with the stress there was a 15 percent increased use of alcohol and an 18 percent increased use of drugs among adolescents and young adults.

WHAT IS THE SOCIAL-EMOTIONAL LEARNING (SEL) BASELINE IN YOUR CLASSROOM?

The teacher needs to focus on working in the classroom and the school to create a positive environment in response to the pressing nonacademic challenge of student social-emotional wellness. It is imperative to recognize that the teacher is ultimately responsible for fostering a positive learning environment anywhere in the school.

What then is a positive environment that is conducive to learning? The environment includes the following four critical elements, and they are (1) a sense of safety, (2) a sense of belonging, (3) a climate of learning support, and (4) fairness of school rules and guidelines.

Let us find out how you can achieve the four elements of a positive classroom by the following teacher self-assessment inventory.

Section (1) Sense of safety

1.1 The teacher maintains a climate of respectfulness in the classroom where teacher and students show consideration for each other.
   (a) seldom   (b) sometimes   (c) always
1.2 The teacher makes students feel safe in the classroom from verbal abuse, teasing, and exclusion from other students.
(a) seldom   (b) sometimes   (c) always
1.3 The teacher has a clear student management plan for the school semester.
(a) seldom   (c) sometimes   (c) always
1.4 The teacher enforces the set student management plan for the school semester.
(a) seldom   (c) sometimes   (c) always

Section (2) Sense of belonging

2.1 The teacher makes the students feel welcome in the classroom.
(a) seldom   (b) sometimes   (c) always
2.2 The teacher is sensitive to the learning needs of students and adjusts his strategies of instruction.
   (a) seldom   (b) sometimes   (c) always
2.3 The teacher finds time to connect with reference to student's extracurricular activities.
   (a) seldom   (b) sometimes   (c) always
2.4 The teacher makes students feel accepted by peers in the classroom.
   (a) seldom(b) sometimes   (c) always

Section (3) Learning support climate

3.1 The teacher gives encouragement and constructive feedback to students.
   (a) seldom   (b) sometimes   (c) always
3.2 The teacher encourages students to think independently.
   (a) seldom   (b) sometimes   (c) always
3.3 The teacher supports students taking calculated risk in learning.

(a) seldom   (b) sometimes   (c) always
3.4 The teacher nurtures a climate of student questioning and discussion.
(a) seldom   (b) sometimes   (c) always

Section (4) Fairness of school rules and guidelines

4.1 The teacher communicates behavior rules clearly to students.
(a) seldom   (b) sometimes   (c) always
4.2 The teacher enforces behavior consequences of rules to students.
(a) seldom   (b) sometimes   (c) always
4.3 The teacher communicates the recommended schedule of the syllabus to students.
(a) seldom   (b) sometimes   (c) always
4.4 The teacher follows the recommended schedule of the syllabus.
(a) seldom (b) sometimes (c) always

Let us add up the survey score to find your efforts to create and support a welcoming class environment of learning. Table 4.3 shows the inventory with four sections 1, 2, 3, and 4. Each section respectively has four items, and each item has three choices (i.e., seldom, sometimes, and always). The selection of "seldom" is worth 1 point, the selection of "sometimes" is worth 2 points, and the selection of "always" is worth 3 points. The total minimal you can score is 16 points and the total maximum is 64 points.

If you score between 16 and 28, you still have a lot to learn and improve. If you score between 29 and 40, you still have work to do for improvement. If you score between 41 and 52, you would expect students to enjoy learning in your classroom. If you score between 53 and 64, you can invite your principal to visit your classroom and present a short class video for the special school of education meeting.

What is your strength in the nonacademic classroom contribution to student success? You can analyze the section subscores individually. Under each section of "sense of safety," "sense of belonging," "learning support climate," and "fairness of school rules and guidelines," the minimal total score is 4 points and the maximum is 12 points. Your efforts are weak if you score between 4 and 6. Your efforts are run-of-the-mill if you score between 7 and 9. Lastly, your efforts are praiseworthy if you score between 10 and 12.

HOW CAN YOU BE THE CHAMPION OF SOCIAL-EMOTIONAL LEARNING (SEL)?

Are you at a point in teaching where you have a solid grasp of the SEL landscape of your students? Things will not move forward on its own if you have the mindset of "I do not need this because my students are okay and I

Table 4.3 Classroom Social-Emotionality Teacher Self-Assessment Inventory

	Seldom (1 point)	Sometimes (2 points)	Always (3 points)
Sense of safety			
1.1			
1.2			
1.3			
1.4			
Subtotal (12 possible maximum points):			
Sense of belonging			
2.1			
2.2			
2.3			
2.4			
Subtotal (12 possible maximum points):			
Climate of support			
3.1			
3.2			
3.3			
3.4			
Subtotal (12 possible maximum points):			
Fairness of rules and regulations			
4.1			
4.2			
4.3			
4.4			
Subtotal (12 possible maximum points):			
Total of four sections (64 possible maximum points):			

Source: Author created.

am doing fine." Instead you need to have the motivation to say "I want to be better because I want my students to reach new heights of success." How then do you imbed SEL in what you do in the classroom? The short answer to the question is that it is all about relationships. Surprise? Let us revisit the four critical elements of a positive learning classroom and find out.

(1) How do you create and develop a climate of safety?

To some students feeling safe can be the security of the school building subjected to the involuntary risk of disasters. School teachers are familiar with the mandatory drill for fire, tornado, and earthquake. What remains not mandatory in many schools and is becoming critical is the drill for school intrusion. Take it upon yourself to inform and practice the school intrusion protocol, and this will be a definite life saver on an ill-fated day that the incident happens. The information and the drill practice imbed in teaching will give students the confidence that school is a safe place to go to everyday.

To many other students being safe means a secure feeling for the teacher and other students. Feeling is psychological and is not tangible; consequently, as a teacher, you can (a) build trust, (b) use more inclusive language, (c) use positive language, and (d) make risk-taking OK for students and for the teacher.

Building trust is the cornerstone of psychological safety. Students will not put their trust on the teacher if they do not feel safe. Practice trust if you want to be the pack leader because it is the currency of good leadership. Do not ask students to do something that you will not do it yourself and that includes giving your trust. Remember one good way to build safety is by placing your validated trust in students.

Placing trust on students is risky business if it is not done prudently. For that reason, validated trust is the flip side of blind trust. As a teacher, you give trust through a process of training, certifying, empowering, and finally trusting.

Training is teaching your student an area of responsibility and is similar to the teacher-centered philosophy of "let me tell you." An example of training can be as simple as giving an assignment with a submission deadline.

Certifying is similar to the student-centered philosophy of "Now you tell me" to validate what your students have learned. Checking to see that the student assignment is done on time is an example of certifying.

Empowering is independent student practice where you give students the authority to make autonomous decisions. Asking students what would be a reasonable timeline for a student assignment is an example of empowering.

Finally, trusting gives the student the full faith and confidence to be on his own. Please note that training-certifying-empowering-trusting are not to be misconstrued as micro-management because it is a sequential process of track record building and validation.

The use of more inclusive language is the next element to build a safe and positive classroom. The way the teacher communicates with reference to the word choice, the tone of voice, and even body language can all be perceived as elements of receptivity or hostility. Compare the following two approaches of a science teacher giving feedback to one of the students:

"Jason, your comment in the evolution versus creationism debate was inappropriate. You spoke over George, cutting him off, and that completely undermined him in front of the debate teams. You were not considerate with what you did and discouraged the productivity of the rest of the discussion by creating an uncomfortable tone."

"Jason, I do not think the comment we made in the evolution versus creationism debate landed the way we intended. It seemed to shut down George and set an uncomfortable tone across the debate team for the rest of the meeting. What was our thought process on why and how that comment was made?"

What is the difference between the two student feedbacks? The first one is quite abrasive using the words "you" three times and "your" one time. It alienates Jason from George and the rest of the class. The second feedback uses the words "we" and "our" to communicate to Jason subtly that although the science teacher gives him constructive feedback, he is still valued, and intends to join him to explore how they together can improve the behavior for the common good of everybody. The second conversation is more inclusive by creating an opportunity for a two-way dialog about the behavior at hand.

Small adjustments in our more inclusive language can have surprising impacts on creating psychological safety for our students. Do you know other avenues that a teacher can pull more inclusive language to make everyone feel safer? Lastly, let us be cautious about the use of inclusive language as long as the integrity of the conversation is not compromised.

The use of positive language is another important tool to build trust and make people feel safe. Ensure you routinely recognize your students for things that should be rewarded, big and small. Small gestures like a personal comment or thank you or a handwritten note can go a long way.

Teachers often say that they are too busy to pay attention to gestures of appreciation. Though true sometimes, but being busy is not a good excuse for not going the extra mile. Make sure that you leverage your time generously and share that with your students. This can be through tutoring, mentorship, coaching, counseling, or other forms of developmental communication.

Teachers can use small discretionary physical gestures within professional, appropriate boundaries to express appreciation for their students. Such gestures can include handshakes or fist-bumps to make people feel safe and appreciated. Remember that these gestures are to be authentic otherwise fakes can in reality do more damage to the relationship at the moment.

(2) How do you create and develop a climate of belonging?

What is the meaning of "belonging" and what impact might it have on students' motivation to learn are what teachers need to find out.

Let us start with what gives minority students the sense of belonging. It may be a representation of faculty and staff from the minority groups. It may be a representation of students' minority group in the books they read and the videos that they watch in class. It may be that there are adult role models for them to follow. It may also be the normalization of the majority culture in education while marginalized subject areas are referred to as ethnic studies. In addition, it may also be the exclusion of minority students in student government or groups and the list continues.

Vice versa what will be the sense of belonging for the majority students if we turn the situation around with an overrepresentation of minority faculty, staff, books, videos, adult role models, and the student government? In essence, a sense of belonging for the student is that he feels included and not alienated when placed in the school environment. Research (Osterman, 2000) shows clearly that academic performance is correlated with feelings of belonging to help them define what is attainable in schools. For that reason, teachers are responsible for a learning environment that makes students feel welcome.

Knowing the demographics and statistics of teachers in the US report helps us understand immediately what is the reality of the education world today (Teacher demographics and statistics in the US, n.d.). The report states that 74.3 percent of all teachers are women, while 25.7 percent are men. The average age of the teacher workforce is forty-two years old. The most common ethnicity of teachers is 73.8 percent white, to be followed by 12 percent Hispanic or Latino, 10.1 percent Black or African American, 3.4 percent Asian, and 0.7 percent American Indian.

With the demographics and statistics report one can teasingly say that the teaching workforce is predominantly white female representation. How do teachers make students feel welcome if they are not a member of the majority group? Here are two suggestions directly related to the teacher.

(a) Change must start with the mind of the teacher. The teacher must be honest and reflective, recognizing his own prejudices and presuppositions about different cultures. Does the teacher set lower academic and social expectation for the minority students? Does the teacher reprimand the minority students more so he can listen to the rest of the class? Can the teacher empathize with all students? Does the teacher really want to make the effort to change how he conducts himself in the classroom? There should be no stereotype based on race, ethnicity, or socioeconomic status, and students of all backgrounds should be welcome with their individualized heritage and culture. Read and learn the history of the country, and you will understand that it is in diversity that the people honestly flourish.

(b) Do you suppose that a positive mind commands positive actions? This is what the teacher can do to be culturally responsive and make everybody feel welcome. Culturally Responsive Teaching (CRT) aims to address a student body with cultural diversity; it makes the effort to welcome all cultures. It gives students the opportunity to take proud ownership of their cultural heritage. CRT is an effort to create an environment where there are no subtle or overt pressures for students to disown their own culture and conform to the majority group culture.

On a sidebar, Critical Race Theory (CRT) is not to be confused with Culturally Responsive Teaching (CRT) because the very sensitive topic might work in just the opposite direction dividing students to be racist, sexist, or the oppressed.

(3) How do you create and develop a culture of learning?

What is a culture of learning, and how can the teacher create and develop one? The short answer to the first part of the question is that a culture of learning is a collection of thinking and working habits resulting in the critical acquisition of knowledge, skills, and attitude. The teacher can create and develop the culture by the following three steps: demonstration, guided learning, and independent student practice.

In the first step, the teacher demonstrates his proven habits of thinking, writing, and conversation to achieve learning. To demonstrate is to show; therefore, the teacher shows the process of thinking aloud which is a form of eavesdropping on someone's thinking. The teacher can do reflective writing which is a form of adding personal reflection to enrich meaning or reinforce interpretation. The teacher can also do metacognitive conversations which teach a student how to think about how they approach learning.

Why are the three modeling examples important? They render the abstract thinking process visible to students. Is this not reinforcing the age-old wisdom that a person thinks and he forgets, a person sees and he remembers?

In the second step, the teacher helps students to work independently by self or in groups after the modeling. Have them visit previous learning ideas, write about the thinking and self-assess their work. Here the teacher as a coach suggests strategies and provides general support.

In the third step, the teacher just gets out of the way. Give students only just enough for them to take off on their own. This can be a debate on a controversial topic, the development of a community project, or any problem that is worth solving. Is this not reinforcing the adage of saying that a person does and he understands? In the example, the teacher is to let students show what they can do. Repeat step one in case that the students just sit around without lifting a finger and just may be the teacher modeling did not get through.

The process of demonstration, guided learning, and independent student practice is not a linear stepped process to develop a culture of learning support. There are situations that the teacher needs to loop back to repeat the steps for better understanding, and for that reason, it is very important that the teacher needs to practice and role model his good qualities of being purposeful, consistent, and patient.

(4) How do you develop and create a culture of fairness?

Do you agree that teachers are judged generally by how much they know and by how much they care? Many veteran teachers will share that everything being equal a teacher is judged more by his care practice of being fair than by his head knowledge of being smart. We can say a person who knows a lot is smart and that is about it! You will be surprised to know that teacher fairness is a lot more than people think because it points to the person's core traits with reference to being ethical, responsible, respectful, trustworthy, and ultimately a human being that matters. What is your answer to this question: "Do I look up to a teacher who is smart or a teacher who is fair?"

Try the following five ideas to develop and create a culture of classroom fairness.

(a) Give every student the opportunity to express themselves. Remember that encouraging a diverse range of opinions makes for a better learning experience for all to include students who are shy to contribute.
(b) Set the class routine to call on all students including those who are reluctant and ask what they think. Setting the calling routine will also make students feel that they are accountable in the classroom.
(c) Ask for a second opinion to check if you are fair. Ask a trusted student or another teacher to observe and give you honest feedback. You may find that you unintentionally spent more time with some students while neglecting others. Please understand that while some students do need additional help, it is just unbiased to give everybody your quality time and attention.
(d) Take the time to work with all students, and let those who are struggling to know that they too have strengths. Awarding students appropriately can make them feel better about themselves as long as you make a point of praising everyone at some point. Avoid power-struggling with students by putting them down because such unreasonable behavior will damage a student's self-esteem. Avoid playing favoritism. Work on removing your bias when it comes to being evenhanded in the classroom. Even if one student works very hard and treats you politely, you just have to be cautious about over-praising the student while ignoring other who also needs your attention.
(e) Grading student work fairly is one of the most important aspects of being a fair teacher. It can be a challenge sometimes to grade students impartially when you already have the assumption who tend to do well and otherwise. Start your grading fresh every time with no influence

from the student's prior performance. Grade an assignment or a test against an established rubric so your grading practice is objective and not based on some subjective whims that you might have.

Is that not true that the discussed four elements of creating or developing a culture of safety, belonging, learning, and fairness are all about people relationship? In this case, the relationships are the connections that are generated between the teacher and the students, which can be spontaneous. When it comes to learning, the various relationship skills are interpersonal verbal and nonverbal communication in addition to building trust, the empathy, and the ability to listen to others to determine the appropriate behaviors to ensure positive human relations.

IS SOCIAL-EMOTIONAL LEARNING (SEL) AN OPTION?

The United States has seen an incredible growth in the prioritization of SEL. Collaborative for Academic, Social, and Emotional Learning (CASEL) scholars learned that three states have state SEL standards that span all grade levels knowing that states play a key role in education. Currently, the three SEL states are Illinois, West Virginia, and Kansas.

The three SEL goals in Illinois are to (I) develop self-awareness and self-management skills to achieve school and life success, (II) use social awareness and interpersonal skills to establish and maintain positive relationships, and (III) demonstrate decision-making skills and responsible behaviors in personal, social, and community contexts. Simply put, the building of positive relationships is the ability to identify, assess, regulate, and control the emotions of oneself, others, and groups.

The substance of goal 1 is the ability to recognize one's emotions and personal qualities with reference to the motivation behind the behavior. The foundation of a person's social-emotional behavior is self-awareness. People who do not know why they are happy or angry are unfortunately at the mercy of their emotions. Positive individuals are able to stay focused on achieving goals.

The substance of goal 2 is to establish the positive relationship through appropriate social skills including the prevention and management of interpersonal conflicts. A person who can control emotions can handle good and bad times properly to bounce back from depression and avoid unnecessary irritability.

Lastly, the substance of goal 3 is about effective decision-making for the well-being of self and others. Here the skill of empathy is key to pick up other people in need of praise or help. Most probably, if you are empathetic you are comfortable with yourself and connected to others.

Find out your social-emotional skills notwithstanding whether you think you know or not sure about them. For our students to learn eagerly and cordially, the attitude shift must first start at the adult level. We as teachers need to know and manage our emotions before we expect to understand and manage the emotions of students. Below is a ten-item social-emotional skills for self-assessment. Give yourself 4 points for each item you select "always," 3 points for "usually," 2 points for "sometimes," 1 point for "rarely," and zero for "never."

(1) I can describe my feelings after an intense social function such as a two-day high school class reunion celebration.
    (a) never   (b) rarely   (c) sometimes   (d) usually   (e) always
(2) I can share how I feel four weeks before the high school graduation.
    (a) never   (b) rarely   (c) sometimes   (d) usually   (e) always
(3) I understand how I feel after leading my debate team to win a scholastic tournament.
    (a) never   (b) rarely   (c) sometimes   (d) usually   (e) always
(4) I can explain how I feel after getting a second place in a school athletic tournament.
    (a) never   (b) rarely   (c) sometimes   (d) usually   (e) always
(5) I persevere in achieving a personal short-term goal that I set for myself.
    (a) never   (b) rarely   (c) sometimes   (d) usually   (e) always
(6) I give no excuses for getting an unsatisfactory grade in a class assignment.
    (a) never   (b) rarely   (c) sometimes   (d) usually   (e) always
(7) I can "see" how my family feel.
    (a) never   (b) rarely   (c) sometimes   (d) usually   (e) always
(8) I pick up subtle signs of what other people want to say in a conversation.
    (a) never   (b) rarely   (c) sometimes   (d) usually   (e) always
(9) My friends feel comfortable with me because of how I make them feel.
    (a) never   (b) rarely   (c) sometimes   (d) usually   (e) always
(10) My friends ask me to lead activities or events because of my interpersonal connections.
    (a) never   (b) rarely   (c) sometimes   (d) usually   (e) always

To find where you stand in the wide range of social-emotional skills you can reference the following scoring guideline. You can be a social-emotional head coach if your score is in the 30–40 range. You can be a social-emotional assistant coach if your score is in the 20–29 range. You can read further on the topic to improve if your score is in the 10–19 range. Finally, if your score is in the 0–9 range, this may be a good area to investigate in great detail if you need a research topic for your term project or thesis.

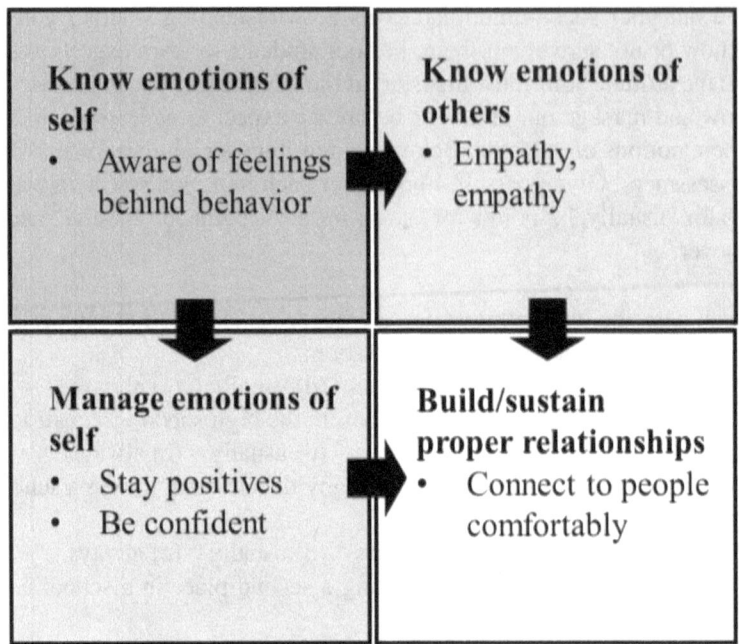

Figure 4.2 The Social Emotional Skills Matrix. *Source*: Author created.

Figure 4.2 shows the connections among the four areas of knowing your own emotions, managing your own emotions, recognizing the emotions of others, and handling/building good and proper relationships.

Daniel Goleman (1995) in his book *Emotional Intelligence* redefines what it means to be smart and argues that when it comes to predicting life's success, emotional intelligence may be a better predictor than academic intelligence. Do you personally agree with Goleman's theory of emotional intelligence reflecting carefully on the ups and downs of your own life?

CHAPTER TAKEAWAY

The wise words of Plato quoted at the beginning of the chapter remind us that while all educational strategies being equal, the viewpoint of the learners should never be overlooked. Effective education is an exceptional combination of art and science because it is impossible to follow an exclusive template to achieve student success.

Researchers and seasoned educators agree that learning and teaching in education are more than academics because of the equally important non-academic foundational support of philosophy and social-emotional wellness.

This pearl of wisdom is applicable to all facets of learning and teaching, and science education is only one example.

Abraham Maslow is right about the human's hierarchy of needs. He states that people are motivated to satisfy certain needs and that some needs take precedence over others. Our basic needs are physiological to sustain life. Once that is fulfilled, the next levels up are the feelings of being safe, love/belonging, acceptance of self, and others. Students are people, and they also have needs. As educators, we do our best to fulfill students' needs through the proper adjustment of value, commitment, and last but not least social emotionality.

Remember for our students to be accepting of one another, the attitude shift must start at the adult level. If we, as the leadership, staff, or teachers, do not display an attitude of inclusiveness and celebrate differences, how can we expect our students to do the same?

REFERENCES

Adler Mortimer biography. (n.d.). Retrieved on May 26, 2022 from https://biography.yourdictionary.com/mortimer-jerome-adler.

Arnn, L. (2022). I said 'Teachers are trained in dumbest parts of dumbest colleges.' Here's why I said it. Retrieved on May 26, 2022 from https://www.yahoo.com/news/said-teachers-trained-dumbest-parts-130116349.html.

Goleman, D. (1995). *Emotional Intelligence: Why It Can Matter More Than IQ.* Bantam.

John Dewey biography. (n.d.). Retrieved on May 26, 2022 from https://www.britannica.com/biography/John-Dewey.

National Alliance on Mental Illness. (2020). Retrieved on May 27, 2022 from https://nami.org/NAMI/media/NAMI-Media/Infographics/NAMI_2020MH_ByTheNumbers_Youth-r.pdf.

Søren Kierkegaard biography. (n.d.). Retrieved on May 26, 2022 from https://exploringyourmind.com/soren-kierkegaard-biography/.

Teacher demographics and statistics in the US. (n.d.). Retrieved on May 27, 2022 from https://www.zippia.com/teacher-jobs/demographics/.

Theodore Brameld biography. (n.d.). Retrieved on May 26, 2022 from https://www.howold.co/person/theodore-brameld/biography.

Toch, T., and Miller, R. (2019). Core lessons: Measuring the social and emotional dimensions of student success. Retrieved on May 27, 2022 from https://www.future-ed.org/wp-content/uploads/2019/02/FutureEd_Core_Report.pdf.

William Bagley biography. (n.d.). Retrieved on May 26, 2022 from https://www.britannica.com/biography/William-Chandler-Bagley.

Chapter 5

What Is the Contemporary Landscape of Science Education?

What we know is a drop and what we do not know is an ocean.

Isaac Newton (1642–1726)

Science is a field which grows continuously with ever expanding frontiers.

John Bardeen (1908–1991)

ANTICIPATORY QUESTIONS

(1) How extensive is the field of science in your modest estimation?
(2) How do you organize the field of science so it will make sense to the students?
(3) How do the process and content of science work together to break new grounds?

A young martial arts student went to a martial art master and said that he wanted to learn the Kung Fu system.

Student: "How long would it take?"
Master replied: "Ten years."
Student: "I am smart, I am ready, and I am willing and able. I want to master it faster than that. I will work very hard and practice what you teach me ten hours every day. How long would it take?"
Master replied: "Twenty years."

One might conclude from the story the simple truth that the more you know the more actually you do not know. Therefore, learning is life long and continuous. The field of science is an excellent example of that truth. The science field is so vast that there is no way that it can be learned, regardless of perspective, with any depth of meaning during any typical school experience. John Bardeen, a history-making two-time Nobel Prize laureate in physics, recognized the expanse of the science frontiers waiting only for the prepared mind to discover. In this chapter, we will attempt to survey the vast landscape of science education.

THE TIMELINE OF MAJOR SCIENCE DISCOVERIES

If we compile everything possible about science, the compilation will come to resemble the thickness of many telephone directories. To be practical, we can only review a tight selection of key science discoveries in table 5.1, table 5.2, table 5.3, table 5.4, and table 5.5. As a guideline, the topic selected generally represents what one can find in many precollege science textbooks.

Table 5.1 Major Science Discoveries in the Seventeenth Century

Date	Major science discovery in the seventeenth century
1600	William Gilbert: Earth's magnetic field
1609	Johannes Kepler: first two laws of planetary motion
1610	Galileo Galilei: telescopic observations
1619	Johannes Kepler: third law of planetary motion
1628	Willebrord Snellius: the Snell's law of light refraction
1628	William Harvey: blood circulation
1638	Galileo Galilei: laws of falling body
1662	Robert Boyle: Boyle's law of ideal gas
1665	Robert Hooke: discovered the cell
1668	Francesco Redi: disproved idea of spontaneous generation
1669	Nicholas Steno: fossils buried in sediment layers of earth
1672	Sir Isaac Newton: white light is a spectrum of colored rays
1675	Anton van Leeuwenhoek: observed microorganisms using simple microscope
1687	Isaac Newton: force of gravitation and the three physical laws of motion

Source: Author created.

Table 5.2 Major Scientific Discoveries in the Eighteenth Century

Date	Major scientific discoveries in the eighteenth century
1735	Carl Linnaeus: system for classifying plants in Systema Naturae
1751	Benjamin Franklin: lightning is electrical energy
1787	Jacques Charles: Charles' law of ideal gas
1789	Antoine Lavoisier: law of conservation of mass, basis for chemistry
1800	Alessandro Volta: invented the battery

Source: Author created.

Table 5.3 Major Scientific Discoveries in the Nineteenth Century

Date	Major scientific discoveries in the nineteenth century
1805	John Dalton: Atomic Theory in chemistry
1820	Hans Christian Ørsted: electromagnetism
1827	Georg Ohm: Ohm's law of electricity
1827	Amedeo Avogadro: Avogadro's gas law
1831	Michael Faraday: electromagnetic induction
1842	Christian Doppler: Doppler effect
1843	James Prescott Joule: Law of Conservation of Energy
1858	Rudolf Virchow: cells can only arise from preexisting cells
1859	Charles Darwin and Alfred Wallace: theory of evolution by natural selection
1861	Louis Pasteur: Germ theory
1864	James Clerk Maxwell: Theory of electromagnetism
1865	Gregor Mendel: Mendel's laws of inheritance, basis for genetics
1869	Dmitri Mendeleev: periodic table of elements
1879	Thomas Alva Edison: the incandescent light bulb
1892	Dmitri Ivanovsky: discovered viruses
1896	Henri Becquerel: discovered radioactivity
1896	Svante Arrhenius: principles of the greenhouse effect
1898	Martinus Beijerinck: viral infection is virus replicating in the host

Source: Author created.

Table 5.4 Major Scientific Discoveries in the Twentieth Century

Date	Major scientific discoveries in the twentieth century
1905	Albert Einstein: theory of special relativity
1912	Alfred Wegener: Continental Drift
1913	Niels Bohr: atom model
1924	Edwin Hubble: Milky Way is just one of many galaxies
1927	Georges Lemaître: Big Bang Theory
1929	Edwin Hubble: Hubble's law of the expanding universe
1929	Alexander Fleming: Penicillin, the disease controlling antibiotic
1932	James Chadwick: discovery of the neutron
1945	Alexander Fleming et al: mass production of penicillin
1947	John Bardeen et al: transistor
1951	George Otto Gey: propagates first cancer cell line, HeLa
1952	Jonas Salk: developed and tested first polio vaccine
1952	Stanley Miller: life building blocks could arise from the primitive earth
1952	Frederick Sanger: proteins are sequences of amino acids
1953	James Watson, Francis Crick: DNA structure, basis for molecular biology
1963	Morley, Vine, and Matthews: paleomagnetic stripes and Plate Tectonics
1996	Roslin Institute: Dolly the sheep was cloned in biotechnology

Source: Author created.

Table 5.5 Major Scientific Discoveries in the Twenty-First Century

Date	Major scientific discoveries in the twenty-first century
2001	The first draft of the Human Genome Project was published
2012	Higgs boson was discovered at CERN
2014	Exotic hadrons were discovered at the LHCb

Source: Author created.

How do you feel after going through more than four hundred years of major science discoveries? Do you feel like staring out the window of a fast-moving train? The five tables highlight the forty-five chronological discoveries in five decades understanding that this is not even close to everything there is in science.

To understand all the items in the long list further we need to do some serious organization. The organization requires that similar items with common properties be placed in the same group and scientists call that the critical skill of classification. After some organization one can draw the following conclusions that science is in reality discoveries and technology with many major divisions.

SCIENCE IS DISCOVERIES

Do you know why the major events of all the five tables are labeled as science discoveries? The word discovery carries the meaning of uncovering information. In the mysterious world of science is the existence of many things, events, and processes patiently waiting for the curious mind. Let us read the following three discovery stories.

The first story is about Anton van Leeuwenhoek (1632–1723) and Robert Hooke (1635–1703). Van Leeuwenhoek (Antonie van Leeuwenhoek biography, n.d.) was a Dutch scientist best known for his pioneering work in microscopy and contributions to the establishment of microbiology as a science discipline. Van Leeuwenhoek was a scientist and a skilled craftsman. He designed and made his own magnifying device. He was curious to observe the detailed structures of very small living things. He often referred to what he examined under the microscope as the wonderful things that God designed in making creatures big and small, and he believed that his discoveries were simply added proof of the marvelous creation. Fascinatingly, Van Leeuwenhoek pursued things that he was curious not to gain fame but to satisfy his craving for knowledge.

Robert Hooke was an esteemed English scholar with broad-ranging learning from astronomy to heat, optics, geology, and paleontology (Robert Hooke biography, n.d.). Hooke published his work documenting experiments he had made with a microscope. In this groundbreaking study, he coined the term "cell" while discussing the structure of cork. He also described flies, feathers, snowflakes, and correctly identified fossils as remnants of once-living things.

The many revelations of the microscopic world were made possible by the imagination of Hooke and Leeuwenhoek in building and using simple microscopes that magnified objects many folds. After the passage of many years, microscopy became the foundations of what we understand the microorganisms in the connection of infectious diseases and the recycling of chemical elements in the biosphere.

One important lesson learned from the van Leeuwenhoek and Hook story is that the use of appropriate tools (i.e., microscopes) is instrumental in the successful discoveries of science. It is intentional that we put the two masters together in the same story not only because they lived around the same time but also because of their groundbreaking collaborative discoveries. Cells exist since the dawn of life, but it is the microscope that helps the naked eye to see more, and it is a significant science discovery.

In time, scientists put their collaborative work together to come up with the Cell Theory. The theory has three premises, and they are (1) all living organisms are made of one or more cells, (2) the cell is the basic unit of structure and function of life, and (3) cells divide to produce new cells so life continues. How can one say even today that the discovery work of van Leeuwenhoek and Hook is not relevant?

The second discovery story is about Isaac Newton (1642–1726). Sir Isaac Newton was a widely recognized English mathematician and physicist. Newton's three laws of motion describe the relationship between the motion of an object and the forces acting on it. The first law states that an object at rest will remain at rest and an object in motion will remain in motion unless there is an external force acting on the object. The second law states that the acceleration of an object equals the net force of an object divided by the mass of the object. If we rearrange the formula in algebra, it will be net force = mass × acceleration. The third law states that forces act in pairs. For every action, there is an equal but opposite reaction.

Newton's laws explain many everyday physical activities including object movement and collision. Basically, the laws affect our lives every time we move or we see something moving. Do you know that Newton's laws of motion and his law of gravity also explain how space objects such as the planet Earth and its Moon orbit around the solar system? The understanding and application of the laws are critical to space travel.

In 1968, the Apollo 8 mission shows that man could successfully do a round trip travel to the Moon. On the return trip back to Earth, astronaut William Anders was asked who was flying the spacecraft. He remarked, "I think that Isaac Newton is doing most of the driving right now" (MetaFilter, 2008). Forces and motion are universally existent, and it is Isaac Newton who painstakingly made the discovery and defined the laws. Yes, Newtonian physics is still relevant today!

Have you seen the magical dance of the aurora borealis in the northern night sky? The third story is about William Gilbert (1544–1603) and the Earth's magnetic field. Gilbert (William Gilbert biography, n.d.) was a medical doctor by training, and he was the court physician of Queen Elizabeth I in England.

Gilbert was most curious about the invisible pushing and pulling of the magnetic phenomena while experimenting with magnet and compass needle (i.e., a small magnet). After years of experiments, he concluded that a compass needle points north–south and dips slightly downward because Earth itself is like a bar magnet with two opposing poles. Behaving like a bar magnet, the Earth has an invisible magnetic field. When a compass needle points to the north, it is actually the south pole of the compass' needle being attracted to the Earth's magnetic north pole. This is how the Earth's north and south poles acquired their names. How many people could meticulously observe and use induction logic to make such a daring encompassing conclusion about the magnetic Earth like Gilbert?

The first person to use the terms electric force and magnetic force, Gilbert is often regarded as the pioneer of electrical studies. Furthermore, he discovered that one could make a magnet out of metals by brushing it with a magnet. Are you surprised that this seemingly simple discovery is modified in today's science classroom to demonstrate the induction of magnetism? In his many tireless experiments, he made the connection between magnetic forces and the rotation of the Earth. His theory that the Earth had its magnetism provided the basis of geomagnetism.

What is the relevance of Gilbert's discovery for us today? The geomagnetic field covering the Earth is the protective umbrella shielding us from harmful cosmic radiation. The aurora borealis or northern lights are the collisions between fast-moving electrons from cosmic radiation and the Earth's geomagnetic field. Additionally, the oxygen and the nitrogen in the atmosphere add beautiful colors to the sky show. Next time when you see the flickering northern lights high in the night sky, you know for sure that you are protected from the deadly solar radiation.

After reading the three stories, we learn that the microscopic organisms, the obscure forces of the moving body, and geomagnetism are all in existence since the beginning of time. It is only through the tireless work of Leeuwenhoek, Hooks, Newton, and Gilbert who discovered their presence and helped us to better understand and appreciate the world around us.

SCIENCE IS TECHNOLOGY

Science and technology are often used interchangeably in lay conversation because of the close relationship. It is important to note that while science and technology are confusingly similar, they are actually different. The main purpose of science is to obtain knowledge of the natural world using the scientific method, whereas the purpose of technology is to develop devices or systems based on the application of science knowledge.

The role of technology is indispensable in many groundbreaking science discoveries for the simple reason that man needs tools to help him solve problems. In the previous stories, we learn that microscopes helped Leeuwenhoek and Hooks to uncover the microscopic world. So, what else is technology?

What do you have in mind when you think of technology? Would you consider the cell phone, a table, and a chair technology? Technologies are devices or systems made by humans to solve problems and to make lives easier. For example, how does Sarah get water from a lake if she lives in a hilltop house? Instead of getting down to the lake to get water, humans invented indoor plumbing to help Sarah pump water from the lake to her house. Here, indoor plumbing is technology. It is a system of devices solving the problem of getting water and making lives easier.

Technologies come in various shapes and forms. Some technologies have wire and plug into the wall. Cell phones help you to get information, play music, take pictures, talk to friends, and more. Other technologies do not have wires whatsoever. An umbrella helps to put you in the shade on a hot sunny day and keeps you dry on a cool rainy day. A pen helps you write down lecture notes, a padlock helps to keep your valuables safe, a fork and a pair of chopsticks help you eat food. Let us go back to our earlier question: "Would you consider your cell phone, a table, and a chair technology?" Now, let us continue with the following two stories about science and technology.

It used to be if you want light you need a fire. Unfortunately, there are problems when you use a flickering flame to light a room because small flames are not very bright, and big flames are not very safe. By the end of 1800s, the race was on to make a better and safer way of lighting using electricity. The first light bulbs did not quite work out because they are too expensive to make, some did not last, and others are not bright enough. Then came the brilliant American inventor Thomas Alva Edison (1847–1931).

The basic idea of the Edison's light bulb is to pass electricity through a thin piece of material called a filament causing it to glow and light up in a glass vacuum bulb (Thomas Edison biography, n.d.). However, the tricky part of the light bulb is to find the right material to glow brightly without burning up.

Edison experimented with numerous materials and failed. Each failure taught him something new making him closer to success. After numerous experiments, he found platinum (melting point of 1,772 degree Celsius or 3,222 degree Fahrenheit) to be the right material for his filament, and the rest is history. Lest we forget, Thomas Edison had accumulated a mind-blowing record of over one thousand patents, or the legal right to control the making, using, and selling of the invention for electric light and power, phonographs, telegraph, storage batteries, and telephone.

Our second story starts with poverty, hunger, and disease as the three big enemies of mankind. Before the discovery of penicillin, there was no effective medicine to completely cure infectious diseases and save lives. Therefore, disease remained the deadliest of the three enemies at that time.

It all started in 1928 when Alexander Fleming, a Scottish physician and microbiologist, while doing his many experiments noticed that there was mold growing in his contaminated Staphylococcus culture dish (Alexander Fleming biography, n.d.). Serendipitously, he spotted a clear area around the mold with no bacteria growth. What happened? After further investigation, Fleming found that the specific mold, called *Penicillium notatum*, could kill bacteria. He was then determined to find a medicine to help people leading him to the discovery of penicillin, the miracle antibiotic drug. The Fleming anecdote is a science discovery that we need to establish before we start the second science and technology story below.

In 1945, the Nobel Prize in Physiology or Medicine was awarded jointly to Sir Alexander Fleming, Ernst Boris Chain, and Sir Howard Walter Florey for the discovery of penicillin and its curative effect in various infectious diseases. The penicillin discovery story starts with Alexander Fleming. However, the mass production of penicillin before it could make a useful medicine is where technology comes into play. The large-scale production of the miracle drug relies on a human-made system involving small-scale seed planting and extraction, large-scale fermentation, separation, purification, field testing, and finally government approval.

The *Penicillium notatum* is the penicillin-producing mold. A logical place to start is to grow the mold naturally with the right raw materials under healthy growing conditions. The production is later increased, transferred, and sustained under the ideal conditions of a laboratory. Imagine a laboratory with many culture apparatuses. To increase the growth volume further, the mold is cultured in deep fermentation tanks with other enhanced nutrient ingredients. At this point of production, the penicillin is mixed with the mold and other manufacturing materials.

For use as a government-approved antibiotic medicine, the penicillin needs to be extracted, purified, and field-tested for final public consumption. How is the mass production of penicillin as an antibiotic medicine technology-based?

Do you remember last time you go to the supermarket to buy breakfast cereal, cheese, canned vegetables, canned meat, bread, pies, ramen noodles, potato chips, bacon, sausage, milk, or orange drink? The food materials are artificially processed transforming them into food through different physical and chemical processes such as mincing, cooking, canning, liquefaction, pickling, macerating, emulsification, adding preservatives, or other food-enhancing ingredients, and more. What is the basis of your decision whether you prefer to squeeze your own orange juice or ready to drink orange juice from a can? Not all processed food is bad because food item such as milk needs pasteurization to make it safe. Will life go back to the prehistoric Stone, Bronze, or Iron Age if there is no technology to help us process food and medicine?

SCIENCE AND ITS MAJOR DIVISIONS

Do you notice that common to the five stories presented is the process of science? Whether it is Leeuwenhoek, Hooks, Newton, Gilbert, Edison, or Fleming, these scientists all attempted to find answers to questions about the world around them, and they were successful. They worked like detectives using various skills to solve complex problems.

In science, trying to find answers to a question is called the scientific method of investigation. Figure 5.1 is an overview of the scientific method of investigation understanding that the order of the steps or the number of steps used in the method may vary. As a result of many investigations, a theory such as the Cell Theory or the Geomagnetic Theory is proposed to explain why something happened. Other times, a law such as the Newton's law of gravity is proposed to describe what happens under a certain situation. Theories and laws are developed from the scientific method of investigation. Please note that the scientific method of investigation is a very important domain in the Next Generation Science Standards as already described at the end of chapter 1.

After reading the science stories earlier in the chapter, it is noted that all the scientists made claims about something that they discovered such as cells, Penicillium, different behaviors of force, the Earth's magnetic field, and the electric light bulb. Claims are statements people believe to be true when answering questions. Let us examine the following claims. All living organisms are made of cells. Penicillium kills germs. The force of an action is the same as the force of the reaction. The Earth behaves like a giant magnet. A platinum filament helps to light up an electrical light bulb.

A claim is worthless if it is not for the support of evidence to show that the claim is true. Evidence is factual information from experiments that contribute to the credibility of the claim. What is the logic that helps to tie the evidence to the claim? The logic is a description of reasoning and/or scientific principles that fit the evidence with the claim. All in all, the claim-evidence-reasoning protocol is a different way to reinforce the scientific method of investigation.

Science with reference to its content and the traditional academic experience is divided into three major branches: physical science, life science, and

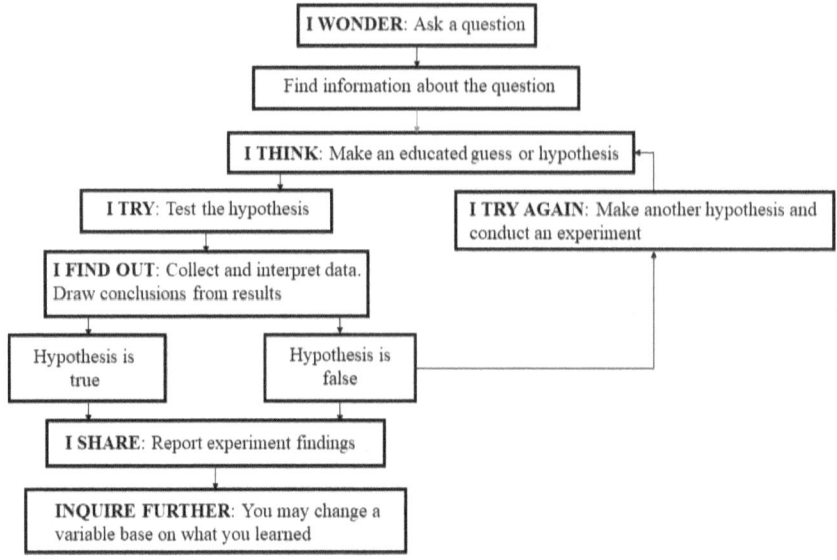

Figure 5.1 The Scientific Method. *Source*: Author created.

earth science. Physical science is the study of matter and energy in courses like chemistry and physics. Life science is the study of living things in courses like biology and microbiology. Finally, earth science is the study of earth and space in courses like geology and meteorology.

To many students, science has the tarnished reputation of being a difficult subject with many unfamiliar vocabularies. One incorrect approach to studying science is vocabulary or definition memorization. Science should be learned and not memorized because the subject is more than a long list of terms. In the hierarchy of learning, memorization without understanding is at the bottom level. In that sense, any science vocabulary not in a context is considered to be weak and will be lost in the unforgiving passage of time. The idea of less is more in learning can be achieved when a large number of vocabularies are grouped into a few concepts. In science education, conceptual learning is proven to be more effective than vocabulary learning.

A long time ago an absent-minded scientist was asked the question "How do you perceive the physical world around you?" Later the scientist wittingly answered, "All things can either be matter or energy." This superficially simple answer puts the physical world into a comprehensive concept. Scientists have discovered some basic concepts to help us understand the world around them. These six concepts are:

(1) All things, living or nonliving, are made of matter. Look around. Every object you can name is made of matter. Matter is anything that has mass and volume including those you can see and other you cannot see. Examples of matter are dogs, cats, paper, water, and air.
(2) Matter is made of very small units called atoms, and they are invisible to the naked eyes. The arrangement and motion of atoms called molecules determine the common states of matter which are solid, liquid, and gas. Solids have a definite shape and volume. Liquids have definite volume but not shape. Finally, gases have no definite shape or volume.
(3) Atoms in matter can interact with each other, and they are active. Other atoms in matter are not reactive, and they are inert. The reactivity of an atoms depends on the structure of the atom with its proton, neutron in the nucleus, and electron(s) going around the nucleus in orbit(s).
(4) Energy has no mass and does not take up space. However, energy can move matter; therefore, energy is the ability to move and do work.
(5) Energy comes in many forms. Sound, light, heat, electricity, and magnetism are different forms of energy. Energy can change forms that are interrelated.
(6) Matter and energy are related. Some matter gives off energy. Other matter uses energy. Without energy, matter will not move and do work. The famous Einstein equation $E = MC^2$ describes the fascinating relationships between mass and energy.

Figure 5.2 shows how the six concepts are reconfigured to form the three major disciplines: physical science, life science, and earth/space science. "All Things" at the top of figure 5.2 represents everything we know with reference to the major science subjects we offer in the school experience. Under "All Things" is the nonliving and the living dichotomy. Going down the nonliving division is chemistry which is the study of matter with examples and physics which is the study of energy also with examples. When the chemistry and physics concepts are merged, they together can explain the many structures and activities of the atmosphere, hydrosphere, and geosphere in earth and space science.

After all, does not the atmosphere, hydrosphere, and geosphere the real-world representation of the gas, liquid, and solid state of matter in chemistry? Does not the burning of fossil fuel and the different motion behaviors the real-world representation of energy transformation in physics? Lastly, the remaining life science branch of figure 5.2 is the study of life at various levels of the taxonomy hierarchy from phylum to species.

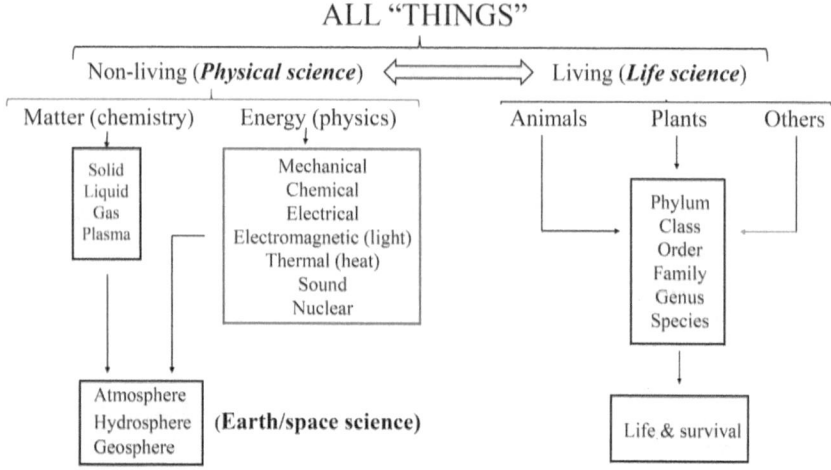

Figure 5.2 Classification of All Things. *Source*: Author created.

One key feature of figure 5.2 is the double-headed arrow between the non-living and the living to underscore the fact that the physical and life sciences are connected. The interdisciplinary relationship explains how the physical world from the use of materials, the application of energy, the climate, and the natural disaster has an impact on the well-being of human life. Do you not agree that any science head knowledge with little human applications will make learning irrelevant and meaningless?

PHYSICAL SCIENCE (CHEMISTRY)

Chemistry is a branch of physical science (Wilbraham et al, 2008). It studies matter and its properties. The heart of studying matter is its small building unit, the atom. Concepts 1, 2, and 3 above are the overarching concepts of chemistry.

The basic structure of the atom of a substance is the neutron (electrically zero charge) and the proton (electrically positive charge) in the nucleus, with the negatively charged electron going around the nucleus in orbit(s). The arrangement of the neutron, proton, and electron described is commonly called the atomic model.

A person can tell the difference between substances by the physical properties and chemical properties. Physical properties of matter can be color, smell, how something can dissolve in another substance, the temperature of boiling and melting, is it attracted to a magnet, and more. The product of any physical change will not change the substance properties because the foundational atomic structure remains the same. Salad is a mixture of different greens that are not chemically bonded. For that reason, the individual green in the salad maintains its properties of taste and color.

On the other hand, substance properties will be different as a result of a chemical change due to the altered composition and atomic structure of the substance. Rust of an iron nail is iron oxide which is the chemical bonding of iron and oxygen. Therefore, the property of rust is different from that of iron or oxygen. Remember, the atomic structure of a substance determines its identity and properties.

It is assuring to know that the physical or chemical change of a substance will not alter the mass of the substance(s) from beginning to end understanding that mass is the amount of matter in a substance and weight is a measure of the gravity pull on the substance.

The law of conservation of mass states that mass in an isolated system can neither be created nor destroyed by physical transformations or chemical reactions. This is an important chemistry concept because scientists will know the identities and quantities of reactants for a particular reaction, and they can also predict the amounts of products that will be produced. Is it not a relief in the science world to know that things just do not vanish because they may be in a different form meaning that the atoms of a substance simply rearrange to form different substances? Now you know why the mass of an iron nail and the mass of the same rusted iron nail are the same.

Any substance in its purest form is an element. When an element is physically combined with another substance, it is a mixture. A mixture keeps the individual properties of the original ingredients such as a salt and pepper mix in a shaker. However, when two elements are chemically combined to form a molecule, the new substance is a compound. The properties of a compound are different from its original substances. For example, the water molecule is a compound. Water is a liquid and is very different from its elements hydrogen gas and oxygen gas.

Why some atoms are reactive to form compounds? The answer is stability. This means that by taking, giving, or sharing electrons with other atoms the resulting molecule will be stable. Will you be surprised to find that achieving the state of stability is the ultimate goal of all activities in physical, life, earth, and space science? You will find out more.

Have you seen scientists mixing bubbly solutions in the laboratory? Solution is a mixture where the dissolved substance called solute is evenly mixed in a solvent, the substance into which the solute dissolves. The strength or concentration of a chemical solution determines the reactivity of a chemical reaction. For example, sports drinks are a solution of sports drink mix powder (the solute) dissolved in water (the solvent). A concentrated solution has a lot of solute while a dilute solution has little solute contained in the same volume of solvent. When you prepare your own sports drink, you will find that the concentration of the solute and the temperature of the solvent are the two factors affecting solubility.

The acidity or the alkalinity of a fluid is an important chemical property of a substance. It involves the breaking or making of a bond containing a hydrogen atom. The property is measured in a quantitative pH (power of hydrogen) scale from one to fourteen. The pH of a solution indicates its concentration of hydrogen ion. As the pH gets closer to fourteen, a solution is more alkaline, and as the number gets closer to one, the solution is more acidic.

The pH of pure water is in the middle of the pH scale which is seven. Human tear has a normal pH range of 6.5 to 7.6, saliva has a pH range of 6.2 to 7.6, and stomach has a pH range of 1 to 5. The above three examples show a very important fact of health that a specific pH range has to be maintained for proper metabolic functions. This piece of information shows one more time how physical science (chemistry) is related closely to life science (biology).

Density is a property behind the sink and float phenomenon. If the density of an object is more than the density of a liquid, the object sinks. Reversely, if the density of an object is less than the density of liquid, the object floats. Density is a ratio between the mass and volume of an object. This explains why a 60,000-ton aircraft carrier floats in water.

Every serious chemistry student needs to know the atomic number and the electron configuration of the elements before he understands and predicts what will be reactive or not reactive in a chemical reaction. Thanks to the ingenious work of Dimitri Mendeleev (1834–1907), a Russian chemist, the classification of the elements was developed (Dmitri Mendeleev biography, n.d.).

Mendeleev found that when all the elements were arranged in order of increasing atomic mass, the resulting table displayed a recurring pattern of properties or periodicity in groups of elements. Figure 5.3 shows that the periodic table has eighteen columns. Fascinatingly, columns 1, 2, 13, 14, 15, 16, 17, and 18 each represents one group or family of elements distinguished by the electron configuration of its orbit(s).

There are eight major groups in the periodic table: 1, 2, 13, 14, 15, 16, 17, and 18. Each group consists of elements with identical electronic structures that are distinguished by filled inner-electron shells, and many electrons are equal to the number of the group in their outermost shells. The elements like lithium, sodium, potassium, and so on found on the left side of the periodic

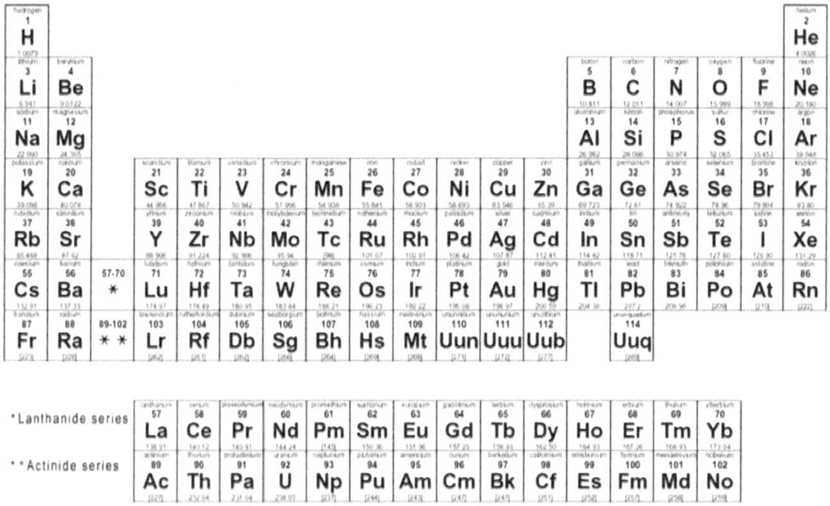

What Is the Contemporary Landscape of Science Education? 115

Figure 5.3   The Periodic Table of Elements. *Source*: Author created.

table are typically metals. While the elements like oxygen, sulfur, and others on the right side of the periodic table are nonmetals.

PHYSICAL SCIENCE (PHYSICS)

"What is energy?" asked a teacher. A student answered, "It is a blob of thing!" This student though far from understanding energy shares a common view of many others in the class. A blob is a difficult shape to describe, and a thing is a generic nondescriptive noun. Energy is a focus study in physics (Kahn, 1996). What then is energy with reference to concepts 4, 5, and 6 we have earlier to describe the world around us? Let us continue.

Energy is a concept and not a thing. Energy is found everywhere, and scientists use the energy concept to explain the many activities found around us. Let us continue with an analogy in finance that we are more familiar with. In finance, people keep track of all the money activities of buying and selling every time they go shopping. As a buyer, the person loses money in exchange for something he needs or wants. As a seller, the person gains money in exchange for something he sells. In the scenario described, there is a shift of wealth in the transactional activity. This concept of wealth can be used to explain how energy works.

In the world and even the universe, every activity that happens is a transaction of energy. In every transactional event, some things lose energy and some things gain energy and the total energy remains the same. The Law of

Conservation of Energy states that energy cannot be created nor destroyed to mean that the total energy of an isolated system always remains constant.

Force is what causes energy to change. It is the buying and selling of the energy transaction. To change your wealth, you buy or sell. To change the energy, you apply force. To further distinguish force from energy, force like velocity, acceleration, and momentum has magnitude and direction. On the other hand, energy like time, volume, and mass has only magnitude and no direction.

Energy can exist in different forms, and they are:

(a) Mechanical energy. It is the energy of motion. It is found in moving objects. Mechanical energy is the sum of two other forms of energy: kinetic energy and potential energy. Motion is everywhere you look even if you do not move. For you to stand still on the ground, your body mass is pulled by the gravitational force. If you stand at the equator, you actually move more than 1,000 miles per hour with the rotating Earth. However, because everything else on Earth is also moving you have no point of reference to see or feel the spin.

What makes a car move and stop is the application of force. Newton's three laws of motion under the earlier "science is discoveries" section describe the motion of all objects in the universe from the moving car to the orbiting planets.

(b) Chemical energy. It is stored in the chemical bonds of compounds. When the bonds are broken, the energy is released and can be converted to other forms of energy. Food, oil, gas, coal, and firewood are good sources of chemical energy. When we eat and digest, the chemical bond of food molecules is broken to release and give us energy. Fat, protein, and carbohydrate are macro food molecules with stored energy.

(c) Electrical energy. It is the energy of moving electrons caused by the interaction of positive and negative charges in matter. When charges in matter interact, they can produce electric forces. When an electric charge builds up on an object such as a doorknob and transfers to another like your hand, it is called static electricity. Lightning is actually an enormous and phenomenal electric transfer or discharge in nature. When electrons or electric charges move, they generate electric current. The lights of your house use electrical energy. The power source like the battery, the conductor like the wire, the load like a light bulb, and the switch are the basic parts of an electrical circuit which can be arranged in series or in parallel.

(d) Electromagnetic energy. It is the energy that travels through space as electrical and magnetic waves. Light is a form of electromagnetic energy. When an object's atoms heat up photons are produced. Light travels in waves like sound and is visible to the human eye. X-ray, microwave, and radio waves also use electromagnetic energy. These waves have different wavelength and frequency. Waves with short wavelength and high frequency like x-ray and gamma rays are high energy, and they are more penetrating.

(e) Thermal or heat energy. It is the energy of moving atoms of matter that affect temperature. An electrical heat coil of an oven uses thermal energy to toast a slice of bread. Thermal energy can be transferred by conduction, convection, and radiation. Conduction is heat transfer from a warmer object to a cooler object like touching a hot stove with your hand. Convection is heat transfer through the movement of air or water/fluid like the magma movement beneath the Earth's crust. Finally, radiation is the heat transfer through electromagnetic waves like the warmth you feel when you sit next to a space heater.

Thermal energy is measured in temperature expressed in Celsius or Fahrenheit. For example, the boiling temperature of water is 100° Celsius or 212° Fahrenheit. The freezing point is 0° Celsius or 32° Fahrenheit.

(f) Sound energy. It is energy from vibration in the matter to transmit sound. For example, talking generates sound waves from the vocal cords which travel by transferring vibration from molecule to molecule and move sound from mouth to ear. Sound waves travel through air, water, and other solid substances, but not a vacuum. The amplitude, wavelength, frequency, and speed are the four features that define the quality of a sound transmission.

(g) Nuclear energy. It is a form of energy stored in the nuclei of atoms. The energy holding the nuclei together can be released by splitting nuclei apart in a controlled manner. Nuclear energy is also released when unstable radioactive nuclei break down. Uranium is an example of an unstable element. Nuclear energy can be used to heat up water, creating steam to turn an electrical generator. Nuclear energy does not pollute the environment, but it does produce toxic nuclear waste.

LIFE SCIENCE (BIOLOGY)

"What is life?" and "How does the first life begin?" are the two big questions raised usually before diving deep into the syllabus of the life science course. The answer to the first question can be found traditionally at the beginning of a biology textbook (Miller & Levine, 2019). The direct answer to the second question is not usually found but can be conjured when the theory of evolution is discussed.

Figure 5.4 shows a big landscape of life science. The picture can be viewed in two parts. The left-hand side shows the chemical foundation of life pointing to the mystery that life is in actuality made from nonliving atoms and molecules. The cell in the center is where the presentation of life begins. On the right-hand side of the picture are the various life functions and activities leading to its continuation.

An interesting takeaway from the picture is the fact that chemistry and biology are intertwined. To show how things work in life science, the person also needs to know the science of energy and that is physics. Figure 5.4 shows that life science is all about life survival and continuation.

Anything living has the life characteristics of growth, takes in and consumes energy, responds to stimuli and reproduction. Do you know that viruses are not classified as living because they have to rely on the host cell to replicate? There are so many living things that they are grouped into a classification of hierarchy from the inclusive kingdom to the exclusive species. The knowledge of life science is focused for the most part at the species level with reference to the grand concept of pattern, structure and function, change, and stability. Is not the concept of classification of living things similar to the classification of elements of the periodic table? Now, we know why classification is such an important skill needed to help scientists to understand the world.

Let us visit the plant kingdom. Plants differ in size from microscopic to huge, but they are all made of cells and many with pigments for photosynthesis. A very common green plant pigment is chlorophyll. One big distinction between plants is those with no seeds and those with seeds. Plants with no

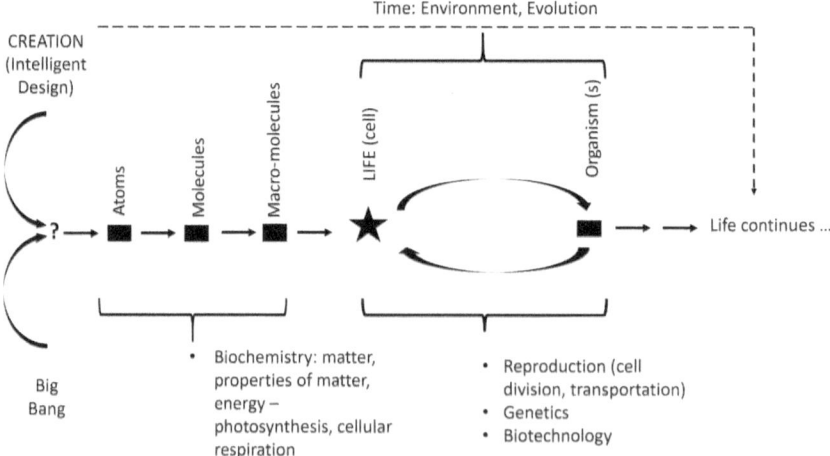

Figure 5.4 The Big Picture of Life Science. *Source*: Author created.

seeds do not have an internal tube system to help them get water and distribute nutrients. Mosses and liverworts belong to this plant category. Plants with no seeds can also have a tube system for water and nutrients transportation. Ferns and horsetails belong to this plant category.

Plants with seeds are the other big classification category. Seeds are reproductive units adapted to land. Seed plants can have seeds that are not protected by fruits. Pine tree, a gymnosperm, is an example. Seed plant, an angiosperm, has seeds inside a fruit. The root-stem-leaf is common to many plants for structure support, transportation of water and nutrients, and for photosynthesis.

The animal kingdom is next, and the two big divisions are invertebrates and vertebrates. Invertebrates have no backbones, and vertebrates have backbones. Sponges, cnidarians, flatworms, roundworms, segmented worms, mollusks, arthropods, arachnids, crustaceans, and echinoderms are all invertebrates. Vertebrates are the other big group of animals with a backbone. Fish, amphibians, reptiles, birds, and mammals are examples of the backbone group.

The human body system is considered the epitome of the life science study because human structures and functions are extremely complex. A good example to illustrate the body's structural hierarchy is to go from the cells to the tissue, organ, and organ system. Is this not all about classification, structure, and function again? Let us visit the ten human body systems.

From the outside, the skin is the largest organ system for injury and infection protection, temperature regulation, sweat excretion, vitamin production in the presence of sunlight, sensing the environment for temperature change, and pain.

The muscular system for movement is next. Muscles that you can control like your arm and leg muscles are voluntary. Muscles that you cannot control like your stomach and heart muscles are involuntary.

The skeletal system is your bones for body support, organ protection, and blood-cell generation.

The digestive system takes in food, breaks it down, and absorbs nutrients into the body. The adult human digestive system is thirty feet or nine meters in length. The tract goes from the mouth to the esophagus, stomach, small intestine, large intestine, rectum and anus. Do you know that the average person cannot stay alive without food for eight to 20 days and without water for three to five days?

The circulatory system is a very special plumbing system with many pipes, the blood vessels and a pump, and the heart. The heart is a kind of involuntary muscle that thumps 3,000,000,000 beats in the average life span. The system moves nutrient and waste in which blood is the transporter.

The respiratory system is what we have for breathing, and it is controlled by the brain and the diaphragm, a muscle. The main organ of breathing is the lungs. We breathe in oxygen which is an important element to help the breakdown of glucose-releasing energy for the body to use. Do you know that the average person cannot stay alive without breathing for three minutes?

The excretory system is a combination of digestive, respiratory, and kidney functions in which the main function is waste removal. Solid excrement as feces is removed through the rectum and anus of the digestive system. Carbon dioxide as a gaseous waste is breathed out by the lungs of the respiratory system. The kidneys filter your blood and get rid of waste and excess water, salt, and minerals. Wastes removal helps the body to maintain homeostasis.

The nervous system collects and relays information about your surroundings to the brain. The brain, in turn, deals with the information and decides on an appropriate response. The brain is the control center, and the spinal cord is the relay of the rest of the body. The brain and the spinal cord make up the central nervous system. All the nerves outside the central nervous system are peripheral, and they can be sensory or motor in function.

The endocrine system is another messaging system using chemicals called hormones. Hormones from endocrine glands like the pituitary, thyroid, pancreas, ovaries, and testes are released via the bloodstream to different parts of the body. Hormones are important chemicals to regulate the sleep cycle, blood sugar, blood pressure, pain relief, and other important metabolic activities. The steady, internal physical and chemical systems are used by all living systems to create balance.

The immune system is an extensive network of vessels, nodes, and ducts that pass through almost all body tissues. The immune system also includes a range of components, including white blood cells (leukocytes), the spleen, the bone marrow, the thymus, the tonsils, adenoids, and appendix. The main function of the lymph system is to defend the body from potentially hazardous microorganisms, such as infections.

The reproductive system allows humans to reproduce sexually so that life may continue. The basic design of the human reproductive system is the male and the female. In the reproductive process, each parent contributes a different sex cell, the sperm and the egg, to create a new cell. This new cell grows and develops and a new human life comes into being. After power-reading the ten human body systems in a whoosh, do you agree that human biology is amazingly complex and organized for the purpose of life and survival?

After the lightning tour of human biology, we need to come back to the base of life—the cell. Cells are the foundational building blocks of life, and the simplest living thing is just a single cell. The cell though small to the naked eye is made of many parts called organelles. Cell organelles with specific structure and functions are nucleus, ribosomes, vacuoles, mitochondria, lysosome, endoplasmic reticulum, Golgi body, cytoplasm, cell membrane, chloroplasts, and cell wall in plants only.

The livelihood of a living organism is made possible by important metabolic activities converting nutrients into energy. The metabolism of a cell includes all of the chemical reactions allowing a cell to survive and thrive. Energy is vital to the survival of all living organisms. Therefore, a logical onset of metabolism is the capture of sunlight energy in photosynthesis. The process converts carbon dioxide and water into chemical energy in the form of glucose and releases oxygen as a waste product.

Put photosynthesis in the reverse is cellular respiration. The process breaks down the chemical bonds of glucose to release the stored energy and liberates carbon dioxide and water as by-products. Cellular respiration is a chemical reaction powered by oxygen. Does cellular respiration help to understand why we inhale oxygen and exhale carbon dioxide?

Protein is an important macro molecule for building cells and tissues. Protein is a string of amino acids, and the sequence of the amino acids determines the protein type. The amino-acid sequence is defined by the Deoxyribose Nucleic Acid (DNA) in the nucleus of the cell. How is it possible that DNA is in the cell's nucleus and protein is made on the ribosome in the cytoplasm outside the nucleus? It can be done through the amazing collaboration of the messenger Ribose Nucleic Acid (mRNA), ribosomal RNA (rRNA), and transfer RNA (tRNA).

Enzymes are important chemicals in many metabolic reactions. Enzymes are protein molecules that help chemical reaction to start and move. Digestive enzymes like your saliva and gastric juice break down food into small molecules while other enzymes help transport small molecules to be absorbed into your digestive system.

Cells constantly move chemicals around in the system. To accomplish the goal of moving are various types of cellular transportation. Diffusion, osmosis, and active transport are the major modes of transportation involving the concentration of the moving molecules and the cell membrane. When transportation is against the high concentration of molecules, energy is needed to push. Among the three, active transport is the one that requires energy to move molecules in and out of a cell.

Cells do not live forever because they die and are replaced. For example, the average life span of a red blood cell is approximately four months versus the average life span of a human sperm is twenty-four hours. New body cells are produced from a nonsexual reproductive cell division process called mitosis. In mitosis, identical cells are produced in a controlled manner called growth. When cells are produced out of control, the cells are cancerous causing the deadly tumor.

In sexual reproduction, sex cells like sperms and egg cells are produced by meiosis and later combined to form the new organism. It is important to note that in mitosis the chromosome number of the daughter cells is identical to the parent cells. In contrast, the chromosome number of the daughter cell is only half the number of the parent cells.

Genetics is a logical next because inheritable characteristics called traits of the organism are passed down and or exchanged from parent to offspring in reproductive cell division. A large part of your visible phenotype such as height, eye, and hair color is determined by the genotypes that you cannot observe.

Gregor Mendel (1822–1844) was considered the father of modern genetics (Gregor Mendel biography, n.d.). He developed the three laws of genetics, and they are the Law of Dominance, the Law of Segregation, and the Law of independent assortment. The study of genetics is of great interest to many because it is connected to many life-changing genetic disorders, diseases, and genetic engineering.

Evolution is a theory developed by Charles Darwin (1809–1882), an English naturalist and biologist (Charles Darwin biography, n.d.). The theory in essence states that species change over time. There are many different ways species change, but most of them can be described by the idea of natural selection. Natural selection is behind the commonly known idea of survival of the fittest. For that matter, a species like the dinosaurs that does not fit its environment become extinct. Together with genetics, the theory of evolution attempts to explain species adaptation and variation. In 1859, Darwin published his monumental work about the origin of species (Darwin, 2022). He, however, did not explain the origin of life.

Ecology studies the interactions between living things and their environment in the confine of an ecosystem. The relationships of organisms are described in a food chain, food web, and food pyramid as producers, primary consumers, secondary consumers, and tertiary consumers. The complexity of an ecosystem goes from the organism to the population, community, ecosystem, biome, and biosphere including all its abiotic and biotic factors. Water, sunlight, food, and living space are the limiting factors affecting the carrying capacity of an ecosystem. Living organisms within an ecosystem live interdependently and competitively through symbiosis, predation, and cooperation.

EARTH AND SPACE SCIENCE

The earth and space science study will be irrelevant and meaningless if it is not connected to the well-being of life. In figure 5.5, the center of the atmosphere, the hydrosphere and the geosphere in the pyramid diagram, is the life's connection called the biosphere. The word sphere in earth and space science is used to denote a large environmental field with its unique components and activities (Lugens, Tarbuck, & Tasa, 2019).

What Is the Contemporary Landscape of Science Education? 123

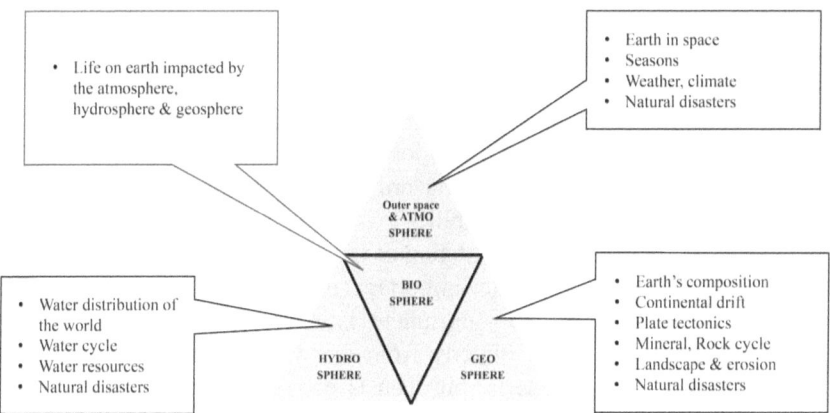

Figure 5.5 The Big Picture of Earth and Space Science. *Source*: Author created.

The solar system has eight planets including Mercury, Venus, Earth, Mars, Jupiter, Saturn, Uranus, and Neptune. Earth is the third rock because of its rocky core from the sun. Earth makes one complete revolution around the sun in 365 days to make a year and rotates on its own axis in twenty-four hours to make a day. The complete revolution of the Moon around the Earth in approximately 30 days to make a month.

The 23.5° tilt of the Earth's axis is the cause of the seasons. Summer occurs on the Earth's hemisphere tilted toward the sun because that side gets direct sunlight. In the meantime, the opposite hemisphere tilted away from the sun receives less direct sunlight and experiences winter.

Planets are space bodies that reflect light. Space bodies that give light are called stars, and the sun is a giant star in the solar system that is closest to Earth. There are other interesting space bodies such as comets, meteors, and meteorites. Not only are these flying objects interesting to watch in the night sky when they enter the Earth's atmosphere but they are the culprits in the Earth's mass extinction events. Is this not a connection to the well-being of the biosphere?

The Earth's atmosphere is a layer of gases about sixty-two miles thick surrounding the planet like the peel of an orange. This thin envelope of gases keeps us warm, gives us oxygen to breathe, protects us from the deadly solar radiation, and it is where our weather happens. Weather is the local atmospheric condition at a given place and time including air temperature, air pressure, wind, humidity, clouds, humidity, and precipitation. Thunderstorms, tornadoes, and hurricanes are severe weather phenomena that can do a lot of damage to human properties and lives.

Climate describes the average weather conditions of an area over a long period of time. Climate comes in different types, and it can be tropical, dry, polar, mild, or continental. A particular climate type of a region is caused by latitude, large bodies of water like ocean and lake, landscapes like mountains, plains, and large cities.

The concept of weather and climate is different, and they should not be confused. Climate change such as global warming, forest fires, and other extreme weather events is a grave concern to many, and how we adapt and act to the change will determine our ability to live on Earth. Weather and climate are closely connected to the well-being of the biosphere.

Atmospheric nitrogen is an important element for building protein. Although 78 percent of the atmosphere is nitrogen, animals and plants unfortunately cannot use it directly from the environment. Thanks to the special nitrogen-fixing bacteria, nitrogen is converted into usable nitrogen compounds. Now plants can absorb them through the roots, and animals can eat plants to get nitrogen. The process described is called the nitrogen cycle. The relationship between nitrogen in the atmosphere and the nitrogen-fixing bacteria is a unique example of the atmosphere-biosphere connection.

All animals are a part of the carbon cycle. When animals eat and digest food, they get carbon from carbohydrates and proteins. In animals, oxygen combines with food in cellular respiration to produce energy and gives off carbon dioxide into the atmosphere as a waste product when animals breathe and exhale. It is amazing to see how the atmospheric components of carbon and oxygen are synthesized through the collaboration of the atmosphere and the biosphere.

All organic matter in you and I is made of carbon from the atmospheric carbon dioxide. In photosynthesis, plants take in carbon dioxide and release oxygen. Conversely, in respiration, organisms breathe in oxygen and release carbon dioxide. This continuous exchange of carbon dioxide and oxygen is called the carbon cycle. The vital exchange of carbon dioxide and oxygen is sustained and balanced thanks to the wonderful work of the carbon cycle. Lastly, carbon dioxide is a greenhouse gas because it traps heat from the sun keeping us warm. However, excess greenhouse gases in the environment cause planet Earth to heat up resulting in ice cap melting, rising sea level, and other disastrous weather conditions.

Getting down from the atmosphere to Earth is the hydrosphere covering approximately 70 percent of the Earth's surface and about 96 percent of it is ocean salt water. Earth is often referred to as the blue planet because of the abundant water on its surface making it the most livable place in the solar system. From that perspective, water is plenty but freshwater for human consumption is scarce. Water is a basic necessity, and no living organism can survive without water. Water is in the atmosphere as water vapor, in lakes and rivers, glaciers and icecaps, in the ground as soil moisture, and in aquifers.

Water resources are used for agricultural, industrial, domestic, recreational, and many other human activities. The wide water usage leads people to ask whether the amount of Earth's water will diminish over time worrying that water as a precious commodity is vulnerable. The short answer is no.

Planet Earth has the same amount of water today as we had hundreds of years ago. How is that possible? Believe it or not that water has simply not been distributed evenly, but thanks to the water cycle the amount of water remains the same. The cycle circulates water between the Earth's oceans, atmosphere, and land, involving precipitation as rain and snow, drainage in streams and rivers, and return to the atmosphere by evaporation, transpiration, and condensation. Is it not rain and snow a representation of the states of matter in chemistry? Is it not evaporation or transportation processes driven by the sun representing the electromagnetic energy in physics?

Let us take note that the problem of water pollution by human activities is a bigger concern than whether we have enough water to go around. Harmful materials from factories, farms, and homes intentionally or unintentionally contaminate our precious water resources. It is therefore critical that water needs to be managed wisely.

Water management is the activity of saving, planning, developing, distributing, and running the optimum use of water resources. Hydropower generation is a way to manage water because it is a good source of energy that is safe, reliable, and renewable with low emissions. Regardless, hydroelectric power also has negative environmental impacts in the dam construction area, and dams are expensive to build.

More than likely you are on solid ground, the geosphere, if you are not flying on a plane or sailing on a ship. Geosphere is about 30 percent of the land surface and splits into large pieces we called continents. Study a world map carefully, and you will see that the western coastline of South America and the eastern coastline of Africa seem to fit together. Is it possible that South America and Africa were together once upon a time was one big piece? Alfred Wegener (1880–1930), a German scientist, proposed the theory of Continental Drift to explain the possibility, but he did not explain how the original huge landmass was broken and pulled apart (Alfred Wegener biography, n.d.). The Earth's internal composition may give us the cue.

Geologists compare the composition of the Earth's structure to the layers of a hard-boiled egg. The thin eggshell is like the outermost layer of the Earth's crust made mostly of rock and soil. Below the crust is the thick layer of the mantle with hot molten rock magma. Going deep below the mantle is the core with extremely hot molten iron and nickel.

Examine a longitudinal section of the Earth's structure. Can you visualize the solid crust floating on top of the fluid mantle? The hot core temperature creates giant convection currents in the mantle to move and shift the crust.

Eventually, the Plate Tectonics theory is proposed to explain how major land masses are moved and created as a result of Earth's subterranean movements. Composition of the earth in conjunction with the movements of the crust explains the formation of mountains, faults, earthquakes, volcanoes, tsunamis, and more.

The solid Earth's crust is made of colorful minerals and rocks. Do you see the presence of these materials around you? How about copper, graphite, or fluoride? Who will not know that diamond is rare to find and expensive to buy. Some minerals in a rock are easy to see because of their sparkling colors.

"Where do rocks come from?" a student asked. The earth science teacher answered quickly, "Rocks come from rocks!" The thinking behind the simple answer is based factually on the Rock Cycle that illustrates how rocks are formed, and they change from one form to another. Rocks may all look the same at first glance, but they are different based on the formation process. The Rock Cycle shows that igneous, sedimentary, and metamorphic rocks are formed differently through heat and pressure, erosion and compacting, melting and cooling.

Fossils are preserved remains or imprints of prehistoric organisms in sedimentary rocks. When a dead organism is buried quickly, the hard parts of the remains are likely to be preserved by mineral replacement, mold, and cast, or occasionally the small original remains preserved in hardened tree resin called amber. Remember the amber cane from a popular dinosaur movie? Some prehistoric organisms such as ammonites lived for a short and specific time period, and they are referred to as index fossil because they provide a reference to date rocks and the history of the Earth's layer which they are found.

History of the Earth is organized by a geologic time scale referencing when certain life forms lived on Earth. Eons, era, periods, and epochs are the major subdivisions of the scale in which eons are the longest time divisions, and epochs are the shortest time division.

Due to the continuous movement of the crust, Earth has not always looked the same. The transformation of the Earth's structure changes not just the appearance but also the organisms living during that time. For example, Mesozoic about 252 to 60 million years ago is the era, and the Jurassic period is the peak when dinosaurs prospered and flourished. The Cenozoic era began about sixty-five million years ago is the era of the mammals. How about humans? We came humbly around 200,000 years ago, so we have not been around very long with reference to the long Earth's history.

CHAPTER TAKEAWAY

The chapter is an ambitious attempt to hit the big ideas in contemporary science leaving many details to underscore the science education approach of less is more. What has been presented is just a tiny drop of knowledge in the ocean. The concepts of science in the context of discoveries and technology are meant to be interdisciplinary because they accentuate the importance of the inquisitive mind, the scientific method of investigation, and the support of using the right tool. The lightning tour of the science disciplines is meant to show the extensive landscape of physical science, life science, and earth and space science not only as an individual subject but more seriously as an integrated science discipline.

REFERENCES

Alexander Fleming biography. (n.d.). Retrieved on June 3, 2022 from https://www.britannica.com/biography/Alexander-Fleming.

Alfred Wegener biography. (n.d.). Retrieved on June 4, 2022 from https://www.britannica.com/biography/Alfred-Wegener.

Antonie van Leeuwenhoek biography. (n.d.). Retrieved on May 20, 2022 from https://www.britannica.com/biography/Antonie-van-Leeuwenhoek.

Charles Darwin biography. (n.d.). Retrieved on June 2, 2022 from https://www.britannica.com/biography/Charles-Darwin.

Darwin, C. (2022). *The Origin of Species*. Ottawa, Canada: East India Publishing Company.

Dmitri Mendeleev biography. (n.d.). Retrieved on June 1, 2022 from https://www.britannica.com/biography/Dmitri-Mendeleev.

Gregor Mendel biography. (n.d.). Retrieved on June 2, 2022 from https://www.biography.com/scientist/gregor-mendel.

Kahn, K. (1996). *Basic Physics: A Self-Teaching Guide*. Wiley and Sons.

Lugens, F., Tarbuck, E., and Tasa, D. (2019). *Foundations of Earth Science*. Pearson.

MetaFilter. (2008). "I think Isaac Newton is doing most of the driving now." Retrieved on June 1, 2022 from https://www.metafilter.com/77688/I-think-Isaac-Newton-is-doing-most-of-the-driving-nowharles.

Miller, K., and Levine, J. (2019). *Biology*. Savvas Learning Company.

Robert Hooke biography. (n.d.). Retrieved on June 1, 2022 from https://www.britannica.com/biography/Robert-Hooke.

Thomas Edison biography. (n.d.). Retrieved on June 3, 2022 from https://www.britannica.com/biography/Thomas-Edison.

Wilbraham, A. C., et al. (2008). *Chemistry*. Prentice Hall.

William Gilbert biography. (n.d.). Retrieved on June 1, 2022 from https://www.famousscientists.org/william-gilbert/.

Chapter 6

How Do You Fill the Cup of Science Learning?

For you the cup isn't half full or half empty, you're always topping it up.

Rowena Cory Daniells

Science is not about status quo. It's about revolution.

Leon M. Lederman

ANTICIPATORY QUESTIONS

(1) What are conceptual learning and teaching?
(2) How is concept mapping a tool for learning and teaching?
(3) What is learning background check?
(4) How do teachers activate students' prior knowledge?
(5) What is the difference between the assimilation and accommodation modes of learning?
(6) What is the similarity between a learning outcome and a learning objective?
(7) How is experiential learning related to progressivism?
(8) What are the key elements of a learning experience?
(9) What are the pros and cons of computer-generated learning?
(10) How did video telecommunication affect the education world during the pandemic years?

THE FLYOVER

A superficial perception of a professional educator is a person with a college degree and a teaching license who talks knowledgeably standing in front of a group. That perception is not accurate because teaching is much more than that. Before entering a classroom, the teacher needs to have a focused idea of what he needs to achieve to help students learn. To a novice teacher, the idea can be a paper lesson plan, to a veteran teacher the idea is all in the head.

The broad idea about what and how to teach typically comes from cross-referencing the curriculum guide, the course syllabus, the textbook, the experience of the teacher, what resources are available, and to finish what is in the final examination. In this view, the teacher is the magician putting all the elements together to make the mysterious concoction of effective learning and teaching. Do you have a teaching toolbox? Take the box with you and start to use the tools that you need to be the magician teacher! In this chapter, you will learn how to use various research-proven tools and strategies for effective science teaching.

CONCEPTUAL LEARNING AND TEACHING

"How do I teach?" is a question that follows the question asking "What do I teach?" How does a teacher teach a Next Generation Science Standards (NGSS) concept now that he has selected one for the lesson plan? The answer to the question is all about the teacher showing how to teach with close reference to the promotion of student learning.

Everything you see around the world however complicated is made of atoms, and they obey the laws and governing principles of science. This structure and function statement is overwhelmingly conceptual. Learning is not a game of petty knowledge acquisition. One effective way to acquire knowledge is to use the broad-stroke approach of conceptual learning.

What is a concept? A veteran teacher defines a concept simply as a big idea with related parts. From the succinct definition, one can further describe the characteristics of a concept as big, general, and inclusive. Are photosynthesis, cellular respiration, density, the laws of motion, the rock cycle, and the water cycle big ideas with related parts? Comparatively, ideas in isolation are weak, and related ideas in a concept are strong. A concept is a condensed version of many connected ideas, and conceptual learning is therefore all about less is more.

What are the components of a big concept in education? This is another way to say that small related ideas constitute a concept. When the learner puts the small ideas in a meaningful manner, it will form a graphical representation that is called a concept map.

Do you realize that graphical representation means visual in the human sense, and it is a dominant human mode of learning? The average human has 770,000 optic nerve cells compared to 160,000 auditory nerve cells to imply the fact that human has forty-eight more times anatomical and physiological support to see than to hear by design. For that reason, one can assume that people in general are more visual learners than auditory learners.

Does not what we just learned explain why a lecture with no visual support is not always the ideal way to learn? The idiomatic expression that a picture is worth a thousand words is used to infer that an image or a visual concept map in our discussion can better describe something than many written or spoken words. For that simple reason, it is always easy to show someone a picture than to tell or ask a person to read about it.

To a student, concept maps are graphical representations of organized ideas to help them see how various topics and processes are connected. To a teacher, concept maps are graphical representations to systematize big ideas in planning or delivering a lesson. Although concept maps come in different forms, the hierarchical concept, the flowchart concept, and the spider concept are the three we will focus below to reinforce the definition that a concept is a big idea with parts, and the parts are related.

Hierarchy concept maps are made by placing a word with arrows to connect it to other words showing the relationships. Some ideas are general, and others are specific; therefore, the parts of a concept map are arranged in a way to form a hierarchical system of relationships. Let us study the following to see how the maps are developed.

The first step to construct a hierarchical concept map is to brainstorm and jot down all the ideas that you can think of related to a concept. Use a word or simple phrase to represent the idea. For example, the three states of matter is a common concept in science and the various related ideas are:

- Water
- H_2O
- Steam
- Ice
- Three states of matter

Second, study the idea list carefully. Which one idea is central to all others? This important idea is the most inclusive and is the one from which all of the other branches out. This is an obvious exercise of classification that requires careful understanding and sorting. Remember, this is a hierarchical map, and the central idea is the hub of all connections that will appear at the top of the map. In the example, the central idea of the list is the phrase "three states of matter."

Third, identify the second most important word(s) from the list. Place the words below the central key word from the second step. Draw an arrow to connect the second most important words to the central key word. In our case, the second most important word is "H_2O," and it is connected to the central phrase "three states of matter."

To finish this part of the map, we need to place a word or short phrase to explain the relationship arrow. The word "of" is used and what we have accomplished thus far is a fundamental unit of the map with two words connected with a proper relationship. The map unit reads "three states of matter" of "H_2O."

Fourth, identify the third most important word (s) from the list. These words will be more specific and should relate back to the words above them, "H_2O" as well as the key central phrase "three states of matter." The third most important words from the list are "Steam," "Water," and "Ice," and they are placed below "H_2O." Connect "Steam," "Water," and "Ice" to "H_2O" with arrows and explain the relationship.

Finally, check the overall organization of the map to make certain that all the words are connected meaningfully. In the three states of matter map, a few more relationship arrows are added to further explain the processes responsible for the state change of solid, liquid, and gas. These processes are "boil," "condense," "freeze," and "melt." The idea relationships of a concept map may vary. For example, one concept can be part of another, it can be used to generate another concept, or there can be a variety of other relationships. Figure 6.1 shows a hierarchical concept map of the three states of matter.

Can you think of other related ideas that can be sorted and organized in a way that is not hierarchical? A sequence of actions or things involved in a complex system or activity is a flowchart concept map. A flowchart is a big idea map with sequential related parts. Similar to the hierarchical map, the construction of a flowchart concept map starts with a key beginning point and subsequently arranges the other ideas in a progressive related manner. Let us study figure 6.2. It is a flowchart concept map regarding how a prepared slide should be viewed properly and safely using the light microscope.

A common mistake of a novice microscope user is to find the specimen of the slide using the high-power magnification objective. The microscope activity has a number of steps, and if not followed properly the user will either not be able to view what he wants to see or damage the slide and worse yet the expensive microscope.

Let us compare figure 6.2 with figure 5.1 in chapter 5. Both are flowchart maps; however, figure 6.2 is a linear map of how to use the microscope, and figure 5.1 is a flowchart map with loops and branches to show the scientific

How Do You Fill the Cup of Science Learning? 133

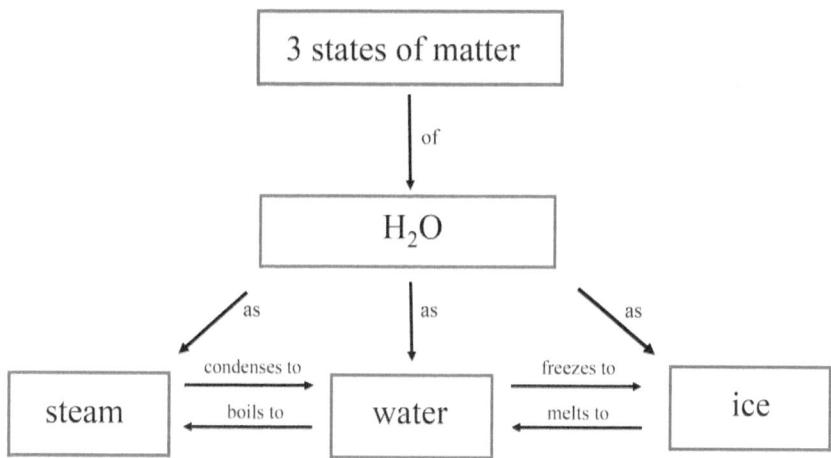

Figure 6.1 Three States of Matter. *Source*: Author created.

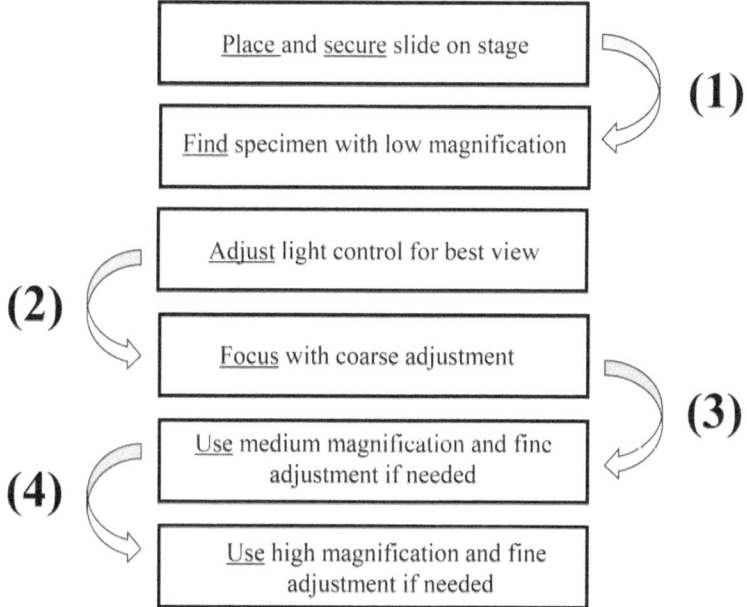

Figure 6.2 Steps of Using a Light Microscope. *Source*: Author created.

methods of investigation. As you become skilled in mapping, you will be creative in your ways of showing ideas and relationships. Let us study the spider concept map next.

The main key idea of a spider concept map starts from the center with the supporting ideas radiating out from the center. The map looks like a spider web hence its name. As a rule, topics that are closer to the center are more important and inclusive. In a map with more than one ring of supporting ideas, the ideas in the outermost ring are less inclusive and less important.

It is just remarkable to compare the ring structure of a spider concept map to the orbital structure of the atomic model. The inner-ring ideas are like the electrons of the first or innermost orbit of the atom; they connect tightly to the center than the outer-ring ideas. Analogously, the strength of the connection to the center represents its importance in the concept.

The first step of drawing the map is the same as before. Brainstorm on a list of ideas surrounding a topic and place the central key idea right in the center of a blank page. Next, identify the secondary important ideas and place them around the center. Connect the central key idea to the secondary idea with an arrow and label the relationship. If there are other ideas that support the secondary ideas then they will be placed in the second ring or even the third and fourth rings from the center. Examples of ideas that support the secondary ideas in the second ring can be specific examples of the concept.

Let us draw a spider concept map of a rock cycle. Here is a list of brainstorming ideas of the rock cycle.

- Rocks
- Heat
- Pressure
- Igneous rocks
- Sedimentary rocks
- Metamorphic rocks
- Weathering
- Erosion
- Melting
- Crystallization
- Compaction

One can dichotomize the eleven rock ideas into two big groups. The first group is rock or rock types, and the second group is all the rock-forming processes. Rocks, igneous rocks, sedimentary rocks, and metamorphic rocks are members of the first group. Heat, pressure, weathering, erosion, melting, crystallization, and compaction are all processes of forming rocks in the second group.

The center encompassing idea of the list is the word "rock" which will be placed in the center of the page. The secondary important ideas that support the center are the three rock types: igneous, sedimentary, and metamorphic.

The three rock types are placed around the center like the first orbit electrons of an atomic model. An arrow is drawn to connect the rock types and the center with a label to show its proper relationship.

To complete the map, one needs to place the rock-forming processes in the right place to explain what the rock cycle is about. Figure 6.3 is a spider concept map of a rock cycle. Do you agree from figure 6.3 that it is accurate to say that rocks come from other rocks and therefore it is a cycle?

As a section summary, we have presented the first critical task of selecting a teaching concept before entering the classroom understanding that the selection is to align it to the appropriate instructional documents including the NGSS.

Drawing a concept map can be as simple as sketching ideas of that on a piece of paper, or it can be even simpler with the help of technology. With the advent of technology, we now have a wide selection of concept mapping applications or apps. These apps are normally free with limited use and are subscription based otherwise. The advantages of using a commercial concept mapping app (figure 6.4) are:

- It is user-friendly because the shape, the line, the arrow, the text, and even the color are all parts of the mapping template. If you can click and drag a computer mouse, you can draw a map in just minutes.
- You can change the shape and size of anything on the map and if you do not, it will be selected and adjusted for you automatically.
- You can change just anything on the map at your fingertips. This easy to change feature encourages the user to try different ideas with ease.

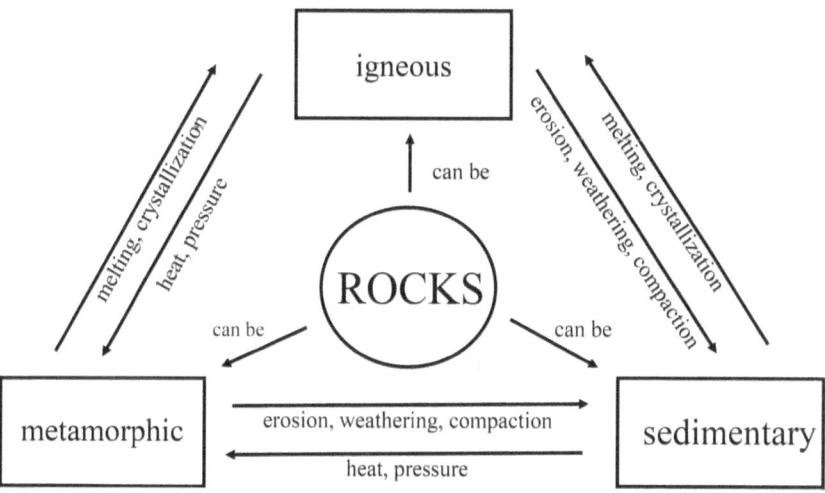

Figure 6.3 The Rock Cycle. *Source*: Author created.

136   *Chapter 6*

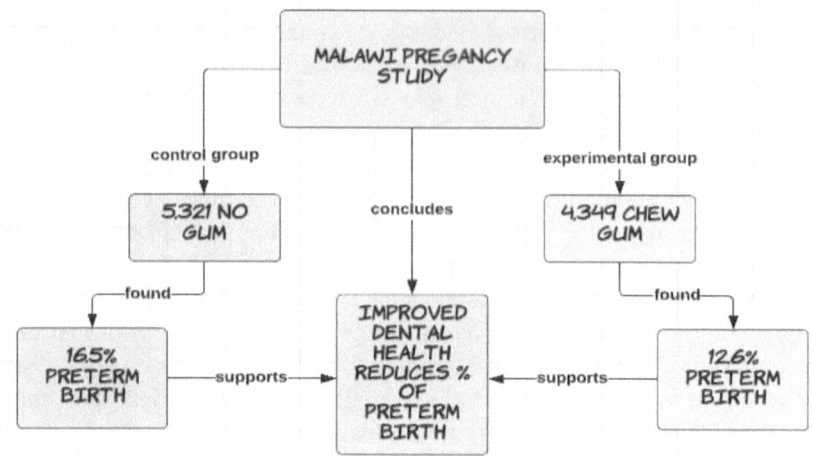

Figure 6.4 **The Malawi Pregnancy Study.** *Source*: Courtesy of Lucidchart.

- Anything you create can be saved electronically. You can start the mapping project and come back days later to finish it.
- Anything you create can be exported or shared by electronic mail attachment.
- If you pick the right mapping app, it might even be free of charge.

Figure 6.4 highlights a scientific research to show that chewing a sugar-free gum with xylitol daily reduced preterm births in a large study in Malawi, Africa (Cunningham, 2022). The article was summarized in figure 6.4 to show that viewing a graphical concept map is a viable option to the alternative of reading a wordy article.

Technology in education is revolutionary, and the concept mapping app is just one example. Technology is another game changer in science education, and we will expand the discussion with more examples in the remaining chapter.

LEARNING OUTCOME

Change is the end result of all true learning.

Leo Buscaglia

How can a teacher be effective when he walks into the classroom with a teaching concept which in reality is just a big collection of ideas? How does the teacher follow through with his teaching concept and transform that into a behavior change or specific learning outcomes that he wants to wrestle next.

In education, learning outcome represents assurance given to students regarding the expectation of a successful learner in a particular course of study. What the teacher expects the student to know and do after the instruction is what defines the course learning outcome. Let us compare the outcome perspective of a student versus the same for a teacher.

A learning outcome statement serves to clearly inform students about what is important knowledge, skills, and attitude in the study. The outcome statement also connects the student to knowledge application in a real-world context, rather than expecting the learner to memorize information. Lest we forget, learning outcome also serves as a reminder for teachers to keep the instruction on track and not to have excessive distraction by trivial information and personal stories.

A comprehensive learning outcome includes a verb to describe an observable action, a description of what the learner will do and under which conditions they will be able to do it, and the performance level they should be able to reach. Learning outcomes can be cognitive, intellectual, verbal, motor skills, or attitude.

Cognitive and intellectual are generally about head knowledge and understanding. Verbal and motor skills are generally about physical ability and performance. Lastly, attitude is a complex internal state that reflects the learner's behavior. Attitude is difficult to measure but can be displayed in the learner's response to people or situations.

From the description above, one gather that measurable outcomes are preferred over non-measurable outcomes to imply that non-measurable outcomes are not as important. This in fact is the shortcoming of not writing anything that is abstract, not observable, and not measurable. Some hard-to-measure outcomes like honesty, diligence, perseverance, team collaboration, and so forth are also very important to the full development of a learner's cognitive and affective well-being.

People often use course learning outcome and course learning objective interchangeably. The difference is the former expresses an observed state to describe what your students actually learned. Whereas the latter is an intended state to describe what you hope your students will learn.

Study the following twelve learner outcome examples to check if the learning expectation is clear to the students. In case that the outcome statement is less than clear, how can you make improvement? In most cases, a learning outcome needs the teacher's verbal explanation to reinforce the fact that the physical teaching part of a teacher is always indispensable.

1. Students will identify situations in which classroom policies apply and describe the proper actions to take in response to them. This type of learning outcome deals with knowledge or intellectual skills. The student understands the new classroom rules that they are being taught.
2. Students will describe the structural difference between a sedimentary rock and an igneous rock by examining a real rock sample. This type of learning outcome deals with knowledge application in the real world. The learner demonstrates the ability to differentiate various rock types.
3. Students will correctly identify the dependent and the independent variable of a plant growth experiment. The learner demonstrates knowledge of what is experimental that can be manipulated and what control that has to be held constant in an experiment.
4. Student will differentiate photosynthesis from cellular respiration with reference to the reactants and products of the complex chemical reaction. The learner demonstrates knowledge of two similar but opposite metabolic processes.
5. Students will interpret parachute-drop experiment data to create a bar graph showing the relationship between the size of the parachute and the weight of the dropped object. The learner demonstrates the ability to do data analysis and graph construction.
6. Students will design a contraption per specification with given materials to protect a raw egg from a twenty twenty-meter vertical drop. The learner demonstrates knowledge application in engineer design in this Science Technology Engineering and Mathematics (STEM) project.
7. Students will determine the dominant eye color of his family by finding the eye color of his siblings, parents, and grandparents. The learner demonstrates his science knowledge application to determine the dominant and recessive traits in genetics.
8. Students will compare the acidity of five given solutions using the pH test paper strip. The learner demonstrates his chemical properties of matter knowledge and identify the solution as acid, base, or neutral.
9. Students will measure the length of a stick shadow in the ground hourly from 8:00 a.m. to 6:00 p.m. The learner demonstrates his knowledge about the relationship between the time of the day and the length of a stick shadow.
10. Students will determine the friction of a flat surface by dragging a given wood block horizontally with a spring balance. The learner demonstrates the knowledge application of measuring friction.
11. Students will count his own pulse for one minute when standing, sitting, running in place, and record the data in a chart. The learner demonstrates the skill of collecting pulse data and correlates the pulse with the person's activity.

12. Students will place a dirty penny in a cup of liquid detergent, soda, and water and determine which solution is the best penny cleaner. The learner demonstrates his understanding of the chemical reaction of copper oxide removal from an old penny.

Do you notice that the start of the twelve learning outcome examples always uses an action verb? The selection of the verb determines the complexity of the outcome statement. For example, one would expect that describing is an easier task than explaining and listing is easier than comparing. For that reason, action verbs used in an outcome statement can be categorized as educational learning objectives based on the levels of complexity and specificity that resemble the shape of a triangle (Bloom et al., 1956).

The lowest category is the recall of facts and basic concepts, and this is represented as the bottom level of a triangle. The first level up from the bottom is the explanation of ideas and concepts representing the process of understanding. The second level from the bottom is application asking the learner to use knowledge in new situations. The third level from the bottom is analysis asking the learner to make connections among ideas. The fourth level from the bottom is evaluation asking the learner to justify a stand or decision. The fifth and top level from the bottom is creation asking the student to produce new and original work.

Where do you think is the cognitive level of concept mapping in the hierarchy? It is analysis, the fourth level from the bottom of the hierarchy.

One easy way to learn the taxonomy is to divide the six levels primarily into two big groups, the low cognitive group in general belongs to asking the what, where, when, and who questions. Contrarily, the high cognitive group in general belongs to asking the how and why questions.

As an education practitioner, what is your questioning habit of asking low level or high level cognitive questions? Do the questions you ask a reflection of your expectation, or an assumption of how the person you communicate with can respond? Regardless of the questioning habits, many veteran teachers like to keep the learning outcomes invisibly in the recess of their mind, others like to write the learning outcomes visibly on the board and use it as a guide of learning for students and a guide of teaching for the teacher.

LEARNING BACKGROUND CHECK

A science teacher is preparing a challenging lesson for the first time in her career. She is confident that she is knowledgeable because she already has taken a number of college courses about the subject matter. The critical question remains finding the right place to begin the lesson. Should the teacher start teaching assuming the students' learning cups are half full or half empty? What can the teacher possibly assume?

Teacher: "We are starting a new lesson this morning, and I would like to find out what you can tell me about photosynthesis?"
Student A: "It is about plants doing something. Is it not?"
Student B: "Is it not like water, carbon dioxide, carbohydrate, and something something . . ."
Student C: "I think it is about green plants making its own food, but how does it work?"
Student D: "I have no clue! What is it?"
Student E: "The whole thing is about capturing energy. Is it not?"
Teacher: "Are you saying that photosynthesis has something to do with water, carbon dioxide, carbohydrate, plants, and energy?"
Student D: "Something like that . . ."

In the above scenario the word "something" is used a few times. In the dictionary something is a word used in various expressions to indicate that a description or understanding being stated is not exact. Going back to the previous classroom scenario: is the cup of science learning half full or half empty? More importantly, what would you suggest regarding how the teacher should introduce the lesson of photosynthesis? Take the question as the first step of "I wonder" of the scientific method of investigation, then we can move forward with collecting more information before answering the question.

Information from the five students tells the teacher that there is a random word list of photosynthesis, and the relationships among the vocabularies are murky. When parts of an idea are missing, it is a lack of knowledge and when the relationship of the parts of an idea is missing the conceptual understanding is weak. As a teacher we need to first find out what students know or do not know before helping them to acquire the designated knowledge and conceptualize the knowledge relationships.

As a school principal, what is the likelihood that you hire John off the street versus hiring Bill whose close friend is your good friend? It is only common sense that you make a better hiring decision about Bill because you can get information about Bill from your good friend about his work and people skills. In that sense, you make decisions based on relational information and not a vacuum.

Background check is a normal process that you have to go through when seeking for employment including applying for a teaching position. There is no typical procedure for background check because organizations may vary when it comes to what they need to know. For example, a car dealer or a real-estate broker may check a buyer's credit, but a fast-food restaurant probably would not.

Schools conduct background check as part of its screening process to guarantee the safety of the students. This procedure is legally mandated in the fifty states of the United States, and it includes verifying academic records, drug screens, and physical examination, and last but not least criminal background record. The purpose of background check is to make the best possible decisions before hiring. Let us do a word substitution and the statement becomes "the goal of learning background check is to make the best decisions possible before teaching."

"The most important single factor influencing learning is what the learner already knows. Ascertain this and teach accordingly" according to Professor David Ausubel (Ausubel, 1968) of Cornell University. When you assume a student's glass of learning is empty, half full, or half empty, you are not doing the due diligence of checking the learning background.

Can one assume that the glass of learning as always full because the visible part of the fill is the liquid and the remaining invisible part is always the air? Can you assume any truth of the statement from your own experience? Expecting all children of the same age to learn the same way is like expecting all children of the same age to wear the same size clothing. The reality of the real world is that students come to us with all kinds of prior knowledge and experience.

When a student is faced with learning new knowledge or skill, the information that the student retrieves involuntarily and immediately from his brain is the existing information that he already has about the subject or remotely about the subject. The existing information is called prior knowledge that is activated and scanned to connect to new learning and problem solving. It is only natural that the learner attempts to connect new learning with old understanding first and foremost.

Students come to learning instinctively with their own prior knowledge, skills, and beliefs. This is a realistic way to say that the glass of learning is hardly ever empty. A person with a deep prior knowledge base is likely to consider an issue from different angles before coming to a response than another person with a shallow knowledge base. Now we better understand why a person might say "no clue" to a situation where he has no prior knowledge or experience. How will prior knowledge influence learning?

Prior knowledge is helpful and sometime not helpful to attain new knowledge. In the worse-case scenario, it might interfere with new learning commonly known as misconception. Do you agree that it is what we already know that often prevents us from learning? A misconception is faulty understanding because it is based on thinking or information that is wrong. Are the following science-related claims facts or fiction and why do you say that?

(1) Diamonds are compressed coal
(2) Vaccines cause autism
(3) Bats are blind
(4) Meteorites are hot when they hit the Earth
(5) Dogs sweat only by salivating
(6) Black holes are literally holes in space
(7) Evolution is a theory, therefore it can be wrong
(8) Venous blood is blue
(9) Microwaves make food radioactive
(10) Planet Earth is closer to the sun in the summer

If you are puzzled by these science claims, you are in good company. Even educated people are divided on the claims depending on their source of information. For school children, their main sources of information are from home, school, and public media. Can you imagine school children learning the wrong information from school and start to share that believing that it is true? In the unfortunate situation that the first impression of learning something is erroneous then the misconception will be difficult to correct according to learning psychology.

What is the source of information about the erroneous claim that in the summer planet Earth is closer to the sun? It is very improbable that school children learn it from their science teacher. Could it be something that they learn it from their own personal experience?

A common personal experience at home is likely to be the toaster, the oven, and the stovetop in the kitchen. When the appliances are on they are hot. When a person gets closer to the hot object, he gets hotter. Is this not a good example of a person's first impression from home? So how can we blame school children for giving us the erroneous claim from their personal experiences? More importantly, how can a science teacher explain the hot summer is related to the part of the Earth that is tilted toward the sun revolving in an elliptical orbit?

The consequence of misconception can be astounding. Many students in our classroom are failing to learn science, and they still hang on to the same misconception that they had when they were in grade school. As a science teacher try asking simple questions in earth and space science about the cause of the seasons and be ready to hear what students have to say. The following are answers from three college graduates (Aguila, 2014).

(1) Seasons happen as the Earth travels around the sun. It gets near the sun which produces warmer weather.
(2) How hot or cold at any given time of the year has to do with the closeness of the Earth to the sun.

(3) The Earth goes around the sun. It gets hotter as we get close to the sun and colder as we get further away from the sun.

What are some suggestions that a science teacher can explain the seasons knowing that some students already come to class with their own misconceptions? Theoretically, new learning is built on the foundation of prior knowledge. Therefore, the more teachers understand what students already know and think, the more they can help students to learn well.

In the real-world classroom, teachers may get the erroneous impression about what students know when they rely on only a small class sample of prior knowledge question and answer. Normally, only a small number of students will volunteer to offer ideas and that poses a challenge to find out what the class already know and how they think. For example, the five students who voice their understanding about photosynthesis are a generous 20 percent representation of the blowing deep space exploration entire class. What about the remaining 80 percent of the class? What if the teacher can introduce a large-scale exploring activity before teaching a new topic? What if this large-scale exploration can be achieved by the use of technology?

Let us explore three approaches to prior knowledge activation below. The approaches are to show a discrepant event, pose a scenario and short answer using technology. The purpose of the approaches is to stimulate the mind and activate the prior knowledge for better learning. In education we call that the bell ringer or mind teaser.

Use a discrepant event at the beginning of teaching a science concept and have students experience an event that is conflicting to what they would expect. The first objective of the discrepant event is, "it does not make sense" striking a cognitive disequilibrium of the learner. Psychologists claim that cognitive disequilibrium or curiosity is an excellent way to motivate learning.

The ensuing discussion after the discrepant event is about "it does make sense now," whereby students ask questions, build upon one another's ideas, and explore each other's thinking. All learning has an emotional base and being curious is one of them. Let us study the bowling ball sink or float scenario below.

A teacher has three bowling balls identical in size and a tank of water on the laboratory demonstration table. "What would you expect if I place the bowling ball one at a time in the tank of water?" the teacher asked. "They will sink" many students responded. "Why is that?" asked the teacher. "Because they are heavy" one student answered. The teacher asked, "Are you a bowler?" The student said, "Yes." The "they will sink" student response represents the existing student's prior knowledge and many speak from the experience that bowling balls are heavy, and heavy objects sink in water.

The teacher picked up the first bowling ball and carefully put it in the tank of water. The ball sank to the bottom. Students cheered and said they were right about their prediction. The teacher picked up the second bowling ball and put it in the water again. The ball floated. Students almost did not believe their eyes and asked the teacher to push the floating ball down. The ball bounced back a few times and floated regardless of how many times it was pushed down. "Let us try the third bowling ball" the teacher suggested. The third bowling ball behaved like the second one, it floated. At the end of the demonstration, the teacher said, "For a bowling ball to sink or float there must be something that is more than just the weight. What do you think? Let us find out."

The bowling ball sink and float discrepant event is about density. The three balls are identical in volume; however, they are different in mass. The first ball that sinks is a sixteen (7,257 gram) pounder, the second ball that floats is a twelve (5,443 gram) pounder, and the last one that also floats is an eight (3,629 gram) pounder.

Any object that floats in water will have a density of less than one gram per cubic centimeter, or one gram per milliliter. As a sink and float discrepant event variation the science teacher can also use the coke versus diet coke sink and float to demonstrate the discrepant event of density.

Do you know that physical science offers numerous attention-getting discrepant event opportunities to introduce or support important concepts and principles? Table 6.1 shows a sample of discrepant events and the related science concept. Please be assured that you will be able to generate your own list of discrepant events as you gain more experience in learning and teaching science.

Presenting a scenario that does not require any specific right or wrong answers is another good way to conduct a learning background check. Many students are hesitant to offer right or wrong ideas due to the potential for public embarrassment. People are likely to get embarrassed when they believe that they have not lived up to what other asks of them or when they are on the receiving end of undesirable attention.

Context also matters, for instance, you probably do not feel embarrassed if you trip in your own house but take that to a party with people watching, and it will be a different story. Therefore, a question just to bounce around ideas should not be intimidating to introduce a concept or warm up a class discussion. Read the following ecology scenario and invite students to make decisions.

A spacecraft is marooned far out in space due to an engine failure. The astronauts radio Earth and ask for help, but they learn that it will take fourteen days

Table 6.1 Discrepant Events and Science Concepts

Discrepant event	Science concept
Air	
• Can you blow up a balloon inside a bottle?	Properties of matter
• Can you blow out a candle behind an obstacle?	Properties of matter
Inertia	
• Can you pull a tablecloth from under plates without breaking them?	Force and motion
• Can you snap a string more than one way by pulling it?	Force and motion
• Can you differentiate a raw egg from a cooked egg?	Force and motion
• Can you filter dirty water with yarn	Force and motion
Density	
• Can you separate fluids into layers?	Density
• Can you sink and float an egg?	Density
Melting, boiling	
• Can you lift an ice cube with a string?	Melting point
• Can you melt ice by applying pressure	Melting point
• Can you boil water without heat?	Boiling point
Potential and kinetic energy	
• Can you make a self-propelled buggy?	Energy
Changes in matter	
• Can you bounce an egg?	Chemical reaction
• Can you write with an invisible ink?	Chemical reaction
Heat	
• Can you have one cold foot and one warm foot at the same time?	Heat transfer
• Why does the first sip of tea always seem hotter than the later ones?	Heat transfer
Light	
• Can you produce infinite images of an object?	Light reflection
• Can you bend light?	Fiber optics
• Can you make a glass jar disappear into thin air	Light refraction
Sound	
• Can you see sound?	Sound property
• Can you make a glass vibrate without touching it?	Sound property

Source: Wong, Ovid (1989). *Is Science Magic?* Childrens Press, Inc. Chicago.

for another spacecraft to reach them with assistance. They check their supplies and find that the air supply is sufficient for fourteen days. They have enough food for ten days and enough water for seven days for normal living. They understand that the more energetic their activities are, the faster their life-support elements will run out. The stranded astronauts must make important decisions about their survival. How can the astronauts establish rules and priorities for conserving resources in such a way that everybody will survive until help arrives?

To conduct the activity in a manner that is effective, small group discussion is recommended. Have students work in small groups and complete a chart with three columns—what they know, what they think they know, and what they need to find out. This can be done on newsprint and reported out to the class. They know and what they think they know are in the domain of prior knowledge.

The recommendations about what to do on a marooned spacecraft will vary according to various priorities and values. From the discussion the teacher will find out the student's prior knowledge regarding what they know about (a) life's requirement of air, food, or water; (b) the importance of resource conservation on the marooned ship compared to resource conservation for planet Earth; and (c) factors in making good decisions.

As an alternative to the marooned spacecraft situation, the science teacher can also ask student to share their life habits with reference to transportation, use of kitchen appliances and home comfort appliances to find whether they are necessity energy user, convenient energy user, or luxury energy user.

With the progression of educational technology, the science teacher can conveniently conduct any technology assistive survey to assess in real time what students already know or for that matter do not know. It will be a challenge to write even a standardized description of assistive educational technology for the simple reason that there are many technology applications in the current market. As a rule, limited use of education technology is usually free to the teacher and any more than that is subscription based. One other indispensable requirement is the student hardware support including iPad, Chromebook, or smartphones and basic infrastructure support from the school. How do we go about using technology to do learning background check? Let us find out.

Educational applications traditionally offer a variety of student pre-assessment options. They may include multiple choice, true or false, or short answer activity. Apparently it is up to the teacher to select the one that he feels is best for student pre-assessment. Short answer is a good choice because the teacher can assess the depth of the student's understanding and the accompanied writing skill to reinforce the importance of language arts integration in science. As another selection, the teacher can instruct students to answer the question eponymously or anonymously. Eponymous is accountability and anonymous is non-threatening.

What is your selection of student pre-assessment and why? Any inclusive science concept if expressed in a simple word or phrase such as density, photosynthesis, or rock cycle can be used as a prompt for knowledge concept pre-assessment. The prompt question can be worded in a way that is non-threatening. In the following examples which one would you choose for pre-assessment and why?

- What is the definition of xyz?
- What do you know about xyz?
- How would you describe or explain xyz to a friend?

Data-driven teaching has never been easier with the arrival of new technology. One nice feature of student pre-assessment using technology is the real-time sharing of information. One can display all the student answers on a screen to view a big picture of what students know or do not know. In some applications, the student data can even be summarized into visual graphs and charts.

Revisiting the science method of investigation, the teacher can role-validate on the spot the presentation of a prompt question, the collection of data which is the student answers, and the analysis and interpretation of the data. Teachers can now use technology to track students' understanding in real time before and throughout the delivery of a lesson or provide students with the results of assigned homework before planning their next lessons.

A learning background check presentation remains incomplete without understanding the psychology of assimilation and accommodation because the effective teacher needs to know the two modes of learning well. Let us start the discussion with a view of your computer desktop with folders. The desktop folders are what people use for quick information storage and retrieval. In that sense, the desktop folders work like the human brain. When you have information that you can place in an existing desktop folder that process is assimilation. What do you do when you have new information that you cannot find and place in an existing folder? You create a new folder for the new information and that process is accommodation.

In learning, assimilation of knowledge occurs when the learners run into a new idea but can add the idea into what they already know. Think of this as placing the new information into an existing desktop folder. Your existing desktop folders are your prior knowledge that you need to activate for storage and retrieval. Are you not convinced now that connecting new knowledge to prior knowledge has its distinct advantage in learning?

On the other hand, the accommodation of knowledge takes more work requiring the learner to create a new space or folder for storage. Think of this as placing the new information into a folder that you have to create and label. Are you more comfortable with assimilation learning or accommodation learning? In learning, assimilation is less challenging than accommodation. For that reason, the ideal lesson should have a good mix of both modes of learning.

As a sectional summary, learning background check with a discrepant event, a scenario, and short answer using technology and more all boil down to data-driven instruction. Learning background check indicates what knowledge and skills students have already learned and in what areas deficiency occurred. This information serves as an important road map for teachers as they plan what to teach next and what areas students might need additional reinforcement.

The learning background check will reveal if a student is advanced, proficient, or basic with reference to the area of learning. Please be cautious that learning proficiency may shift from one area to the next over time, and teachers are not to pigeonhole student's academic capability based on one or two assessments. Equipped with the student information, now teachers really get to understand why one class might not be progressing as fast as another in the same subject at the same grade level.

The background check information will also have teachers set up instructional adjustment for students. For example, a teacher may move students at the basic level to the front of the classroom so they have easy access to additional support. On the other hand, the teacher may provide advanced students with alternative activities that offer them a greater challenge for learning.

Educators need to use a mix of pre-assessment methods to evaluate student readiness. Using one method in isolation misses other opportunities to get information about students' strengths, weaknesses, and preferences. When all is said and done, learning background check is to get a quick informal dashboard reading on student's basic comprehension and engagement.

THE LEARNING EXPERIENCE

All genuine learning comes through experience.

John Dewey

Learning experience is purposeful, and sometime serendipitous interaction lead to learning. In science education, the learning experience can be offered in many different traditional and nontraditional settings such as lecture, experiment, debate, discussion, special field trip, technology-based learning, and the long list continues.

From the previous education philosophy discussion, we learn that hands-on and minds-on are good ways to learn science. Hands-on is collecting, experimenting, constructing, designing, and other physical type of engagement. Minds-on is reflecting, conceptualizing, and other mental type of engagement. Therefore, hands-on and minds-on learning in tandem is doing and thinking period. It is used often as a catch-all expression for educational activities that are self-motivated, engaged, relevant, and experiential.

In typical hands-on, minds-on activities, the learner encounters a number of physical and mental activities. Physically, the concrete experience can be a well-designed computer simulation that we see so commonly used nowadays in distance virtual learning. Mentally, the learner assimilates or accommodates the experience internally in conjunction with his acquired knowledge or skills. The new acquisition is then conceptualized and transformed to new learning. Finally, the learner applies the new knowledge and skills to the real work world for assessment and refinement. Please understand that the description does not imply a tight single file event sequence.

In the context of hands-on and minds-on learning, a revolutionary example is special field trip. Field trip is studying beyond the four walls of the classroom in the traditional sense. How can students not cherish the memory of visiting the nature preserve, animal rescues, manufacturing plant, museums, hospitals, and so on? Do you remember seeing groups of students in the museum moving from exhibits to exhibits and the main activity is observing, writing, and reporting. Does the museum field trip encourage the young minds to have learning interactions? This museum kind of field trip meets the requirement but does it align well with the definition of experiential learning? May be not.

Let us visit Fermi National Accelerator Laboratory (Fermilab), located just outside Batavia, Illinois, near Chicago (n.d.). The purpose of sharing the Fermilab experience is for educators to better familiarize experiential learning in the real world and to make the most out of similar facilities in your area.

Why is Fermilab unique? Fermilab is not just a world-class government research center, a museum, a learning center, free and open to the public, Fermilab is all of the above!

Fermilab is a world-class government research center and is America's premier particle physics and accelerator laboratory. What secrets do the smallest, most elemental particles of matter hold? What is the nature of dark matter and dark energy? How can they help us understand the intricacies of space and time? How did the universe begin? These are questions that Fermilab research scientists attempt to find answers for the benefit of mankind. Is this not similar to the first step of "I wonder?" of the scientific method of investigation we discussed in the previous chapter?

How do Fermilab scientists conduct investigative experiments to find answers? Well, they have very expensive equipment to work with. Fermilab is the home of the circular particle accelerator which is currently the second largest and highest-energy particle accelerator in the world. In the experiments, a high-speed high-energy particle beam is generated and made to collide with a target. The particles and radiation from the collision are then recorded for analysis, and this is how high-energy physicists study the elemental particles of matter. Is this not similar to the "I try," "I find out," and "I try again" steps of the scientific method of investigation we discussed in the previous chapter?

Fermilab has a dedicated office of technical publication for public reports written by Fermi research employees including preprints, conference presentations, theses, and other reports. Is this not similar to the "I share" step of the scientific method of investigation we discussed in the previous chapter? Talk to the scientists, visit the facilities, and read the publication to get your personal experience of what the real-world scientific method of investigation is about.

Fermilab has a few Nobel laureates. Leon Lederman was the second director and the trailblazing physicist who received the 1988 Nobel laureate in physics. Lederman was passionate about making complicated science understandable to precollege students. Is this not what effective science teaching is about? Lederman transformed his passion for science education to the opening of the Science Education Center on Fermilab campus for students and teachers. The Lederman Science Center is the next stop of our special field trip.

The Science center is a stand-alone building away from the main campus facilities and offices. The main attraction of the center is the hands-on ideas room and the interactive science exhibits. The ideas room shows that there is a magical connection between the biggest and the smallest things in the universe, and the laws of physics are shockingly applicable to the immense universe through the endless passage of time.

The interactive exhibits in the ideas room introduce visitors to the basic concepts behind Fermilab's science, and they are the scale of the universe, particles and force, symmetry in nature, and particle astrophysics. The ideas might look arrogant at first sight but the hands-on activities (figure 6.5 and figure 6.6) transform abstract science concepts to concrete learning. This is the power of hands-on, minds-on learning.

The hands-on exhibits are impressive because they are user-friendly equipments that help to understand complex problems. Here the visitor can discover the tools and methods scientists use to explore science. The following is a short sample list of questions from individual science center exhibits.

1. How is light reflection like the bounce of a ball?
2. How do you use motion to find what you cannot see?
3. What do you learn about subatomic particles by analyzing collisions of accelerated particles and target particle?
4. How do scientists use the scattered trails of particles to determine the interactions of particles in a collision experiment?
5. How do you find the best protection material for radioactivity?

How Do You Fill the Cup of Science Learning? 151

Figure 6.5 Fermilab Science Education Center. *Source*: Courtesy of Reidan Hahn, Fermilab.

Read through the five questions carefully, and you will find a pattern of going from simple to complex investigation and application reflecting the spirit of the Fermilab scientific community. As one visiting education group expressed that it is impossible to provide the culture, the equipment, the expertise of the professional staff, and everything in the real world of science and transform it in a way to help the public understand and appreciate. In essence this is what entails the special field trip. There might not be a Fermilab in your area; nevertheless, find a similar institution that provides and stresses the hands-on, minds-on mode of science learning.

Even before the COVID-19 pandemic, educational technology has been an indispensable part of learning and teaching. Here we want to discuss educational technology as another game changer in science education. An award-winning science teacher in the 1990s remembered using email to work with a colleague two thousand miles away in a collaborative acid rain project and that was considered cutting edge before other progressive computer hard and software came along. Technology can be many things, but educational technology focuses on the tools and media that assist in the communication and development of knowledge, and the productivity of both students and teachers.

152   *Chapter 6*

Figure 6.6 Fermilab Science Education Center. *Source*: Courtesy of Reidan Hahn, Fermilab.

What is the status of technology relevance in education today? The COVID-19 pandemic demonstrates one more time bigger and louder that technology is a vital part of learning and teaching. Let us continue.

Student exploration is an indispensable part of the science lesson with reference to hand-on and minds-on learning per our earlier discussion. Due to the lack of proper funding support and other legitimate reasons, the "wet lab" experience is becoming less frequent and more cherished. The use of technology is not meant to be a substitute for proper first-hand experience. Nevertheless, it can be used when the suitable situation arises to provide the virtual learning experience. Let us find out from figure 6.7 how a viable computer-generated science simulation can be used as an alternate student science exploration.

EXPLORE LEARNING

Figure 6.7 is a one-of-a-kind experience to study the conditions of plant growth and for student to understand the often confused dependent and independent variable concept. When the teacher selects to conduct the actual plant growth, he is prepared to do it for at least thirty days with a space, usually by

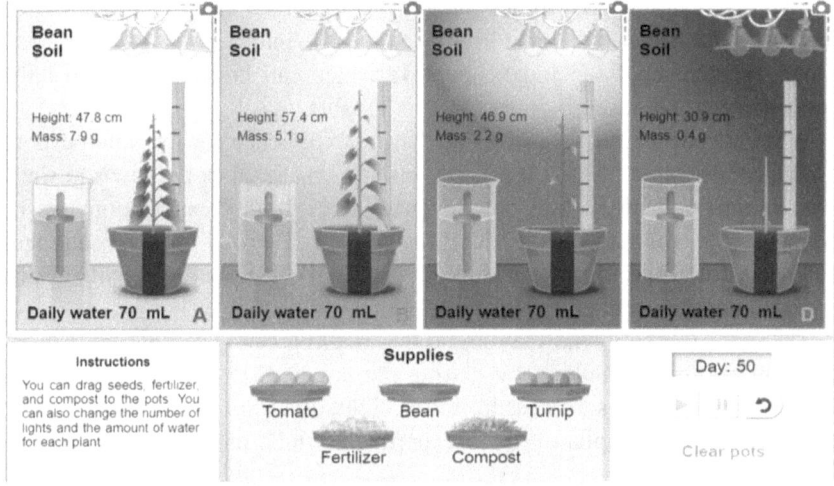

Figure 6.7 Plant Growth Experiment Simulation. *Source*: Courtesy of ExploreLearning Gizmo, Plant Growth.

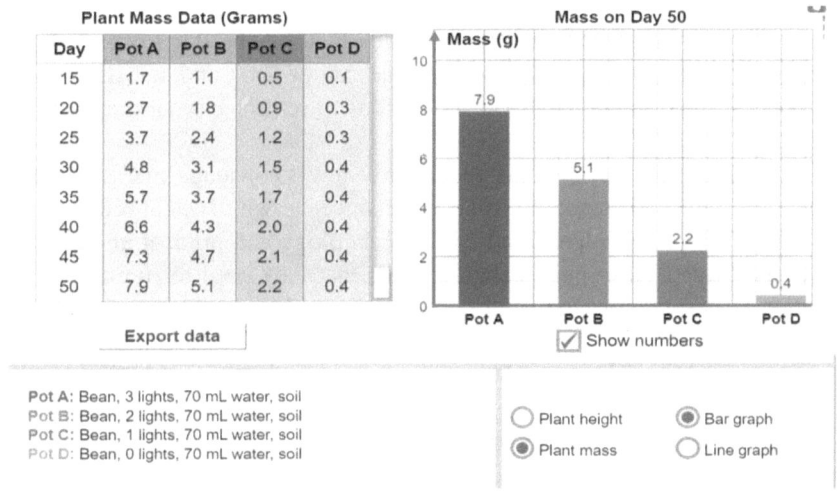

Figure 6.8 Plant Growth Experiment Data. *Source*: Courtesy of ExploreLearning Gizmo, Plant Growth.

the window sill, soil, pots, seeds, and good light-water-temperature control. In figure 6.7, students can easily work with all the virtual supplies and equipment with no fuss for a fifty-day duration. The experimenter can choose from different seeds to compare the growth with reference to plant height or plant mass.

Figure 6.8 shows how the dependent variable data can be measured and recorded in a table and transformed seamlessly into a graph. At the same time, the change of the independent variables can be viewed against the dependent variable clearly in a comparison table.

What are the advantages of using virtual experience for the sole purpose of student learning? Many teachers will say that it all boils down to time, preparation, and instruction convenience. A thirty-day exploration can be condensed to a duration of a few class periods. There is no mess, and students are not afraid to make mistakes because the experience can always be repeated at one's convenience.

There are teachers who may argue that nothing can substitute the experience of learning something by getting the hands dirty with seeds and soil and using real devices to measure water, light, and time. If you have the two options of doing the plant growth experiment which one would you choose and why?

The effective use of digital tools is beneficial to both students and teachers in and outside the classrooms. To the students, it enhances learning engagement because the curiosity to learn through the use of well-designed technology helps students to get a better understanding of science concepts. Technology offers students with information that they need to know with the click of the computer mouse, learning that can be 24/7 and self-paced, and most of all pleasurable hands-on simulations to reinforce what they need to learn.

Technology entices students to explore new knowledge and skills and facilitates their understanding of abstract concepts, with special reference to STEM. Through the appropriate use of technology and prudent guidance of teachers, students also gain technical skills that they need for future jobs in the twenty first twenty-first-century real work world.

To the teachers, educational technology helps them to be creative in their lesson design and delivery and simplifies work productivity. In other words, it enables teachers to improve their methods of instruction and personalize student learning management. Many veteran teachers remember going through the teaching folders and student records in the file-packed cabinet to access information to prepare lessons. Those days are gone, and technology is quickly taking over.

Today, teachers can leverage technology to achieve new heights of productivity. Lest we forget what is good for student learning can also be something that teachers should consider for teaching and technology can be that edge. Let us analyze figure 6.9 to find out how technology can help with learning and teaching.

The centerpiece of the science lesson starts from the NGSS concept in alignment with the curriculum and syllabus documents. In the figure, there

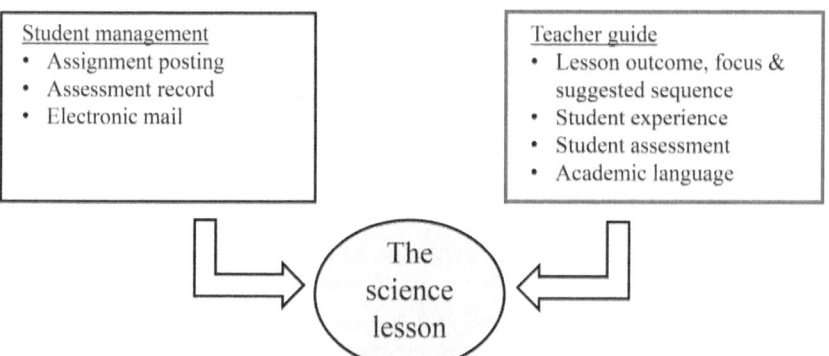

Figure 6.9   The Science Lesson with Technology Support. *Source*: Author created.

are two major components that feed into the science lesson, and they are the teacher guide and student management. If the teacher uses the traditional textbook program, he might have to access various locations to find all the available resources. What if the teacher can find all the resources in an app at the taps of the computer keyboard? In the plant growth example discussed, under the teacher guide you can conveniently find the essential pieces such as the lesson outcome, the lesson focus, the suggested lesson sequence, the student experience with relevant academic language, and the end of the unit student assessment.

At this juncture, it is important to mention that the science academic language or vocabulary can be taught directly or indirectly as long as we understand its impact on learning. Science vocabulary introduced at the beginning of the lesson is feeble in context because it is domain-specific that is beyond the realm of high frequency and basic words. Be careful that without proper context support, many students might take vocabulary learning as mere senseless memorization. On the other hand, science vocabulary introduced in the context of the lesson concept is direct learning. Science vocabularies need to be learned and applied correctly in relevant conversation and report writing.

Student management with reference to accurate record keeping is normally a drag for teachers. This can mean tons of assignments, record keeping, and student performance assessment. "Would it not be a dream to just teach with no assignments or test to administer or student record to keep?" This is apparently a teacher's dream. Fortunately, the paperwork can actually be alleviated by the use of technology. For that reason, the challenging student management part of a teacher's job with the help of technology can now just be a nightmare of the past.

This learning experience section of the chapter is incomplete without an assessment of video telecommunication. The world of education learned how to use video telecommunication overnight in late 2019 when the COVID-19 pandemic hit. The decision was tough because many schools were not prepared to make the switch. Students and teachers alike had no choice but to learn a brand new way to learn and teach. If it is not for video telecommunication schools around the United States, learning would come to a screeching halt and valuable time of learning will be lost.

We heard from stakeholders in the school community that online school is not as effective, and people are getting tired of not getting person-to-person interaction. Can you imagine the grave consequence of schools meeting in person during the pandemic years and adding to the total accumulative death toll of close to a million and confirmed cases of over a million in the United States? People often say that nothing is perfect and that is not meant to be used as an excuse. What we confront in education is always challenging, and if this were an experiment, think of the many variables that we have to juggle to find answers.

CHAPTER TAKEAWAY

The half-filled glass of learning is a philosophy to remind educators that teachers are not to assume the learner's prior knowledge and experience. Little or no knowledge of what a student knows will make teaching like shooting from the hip or worst yet like shooting in the dark.

Students learn better from closely knitted ideas than from a collection of scattered ideas; therefore, conceptual teaching should be an important part of the teacher's toolbox. From the teacher's perspective, the use of conceptual teaching has a better chance of learning success. Many students are visual learners and different graphic organizers including concept maps are recommended to help them better learn and express themselves. Hand-drawn or computer-generated concept maps are used to explain complicated ideas, and it is yet another important assessment tool in the teacher's toolbox.

In the classroom, the teacher is to be cognizant about the right mix of learning assimilation and accommodation so as not to make learning too easy or too difficult. The student learning experience is what the teacher needs to plan and deliver meticulously to align with the students' learning needs, the learning standards, the curriculum, and the course syllabus. Believe it or not, this experience is what will or will not ignite the fire of learning.

In education, we have entered a new era of technology. Can educational technology fulfill all the purposeful activities directed at achieving knowledge, skills, and other desirable character traits? The short answer is no. Technology is only a tool and not an end in itself. We need to keep in mind that the promise of educational technology lies in what teachers do with it and how it is used to best support students' learning needs.

REFERENCES

Aguila, D. (2014). Misconceptions for seasons—Harvard grads. Retrieved on February 2, 2022 from https://www.bing.com/videos/search?q=youtube+misconception+of+harvard+graduate+about+the+four+seasons&view=detail&mid=6E2C7E1C1EDE76837B2A6E2C7E1C1EDE76837B2A&FORM=VIRE.

Ausubel, D. (1968). *Educational Psychology: A Cognitive View*. New York: Holt, Rinehart & Winston.

Bloom, B. S., Engelhart, M. D., Furst, E. J., Hill, W. H., and Krathwohl, D. R. (1956). *Taxonomy of Educational Objectives: The Classification of Educational Goals. Vol. Handbook I: Cognitive Domain*. New York: David McKay Company.

Cunningham, A. (2022). Chewing sugar-free gum reduced preterm births in a large study. Retrieved on February 5, 2022 from https://www.sciencenews.org/article/preterm-birth-chewing-gum-sugar-free-oral-health-malawi.

Fermilab. (n,d.). https://www.fnal.gov/.

Wong, O. (1989). *Is Science Magic?* Chicago: Childrens Press, Inc.

Acronyms

AAAS	American Association for the Advancement of Science
ACI	American Competitiveness Initiative
ACT	American College Testing
AMI	Alliance of Mental Illness
ARPA	Advanced Research Projects Agency
ASCB	American Society for Cell Biology
ASEL	Academic Social Emotional Learning
BSCS	Biological Science Curriculum Study
CASEL	Collaborative Social and Emotional Learning
CEMS	Chemical Education Materials Study
CCS	Common Core Standards
CCSS	Common Core State Standards
CERN	Conseil European por la Recherche Nucleaire
COVID-19	Corona Virus Disease 2019
CRT	Critical Race Theory
CRT	Culturally Responsive Teaching
DNA	Deoxyribose Nucleic Acid
DOD	Department of Defense
GMO	Genetically Modified Organism
GPA	Grade Point Average
IAEEA	International Association for the Evaluation of Educational Achievement
IAU	International Astronomical Union
IHE	Institutions of Higher Education
ESCP	Earth Science Curriculum Project
ISBE	Illinois State Board of Education
IGAP	Illinois Goals and Assessment Program

ILE	Illinois Learning Standards
ISA	Illinois Science Assessment
ISAT	Illinois Standards and Assessment Test
IST	Illinois Science Test
IPS	Introductory Physical Science
ISCS	Intermediate Science Curriculum Study
LEA	Local Education Agency
MAST	Master of Arts in Science Teaching
MSP	Mathematics Science Partnership
NABT	National Association of Biology Teacher
NAEP	National Assessment of Education Progress
NASA	National Aeronautics and Space Administration
NCES	National Center for Education Statistics
NCLB	No Child Left Behind
NGSS	Next Generation Science Standards
NDEA	National Defense Education Act
NSF	National Science Foundation
OECD	Organization for Economic Cooperation and Development
PISA	Program for International Student Assessment
PSSC	Physical Science Study Committee
RFP	Request For Proposal
ROE	Regional Office of Education
SAT	Scholastic Aptitude Test
SBOE	State Board of Education
SEL	Social Emotional Learning
STEM	Science, Technology, Engineering, and Mathematics
TAP	Tapping America's Potential
TIMSS	Trends in International Mathematics and Science Study
USED	US Department of Education
WHO	Word Health Organization

About the Author

Ovid K. Wong is a science education professor at Benedictine University in Lisle, Illinois. He received his B.Sc. and Dip.Ed. from the University of Alberta, his M.Ed. from the University of Washington, and his Ph.D. in curriculum and instruction from the University of Illinois. His public education work experience spans from the inner-city classroom of Chicago to the suburban office of the assistant superintendent. He is the 1989 recipient of the National Science Foundation's Outstanding Science Teacher in Illinois award and the National Science Teaching Achievement Recognition (STAR) award from the National Science Teacher Association. He was the first recipient of the outstanding alumni award from the University of Alberta in 1992 and the first recipient of the distinguished alumni award from the College of Education at the University of Illinois in 1995. In 2013, he received the Distinguished Faculty Award in recognition of his significant achievements in research at Benedictine University. Ovid is the author of thirty-four books and a fervent table tennis player.

www.ingramcontent.com/pod-product-compliance
Lightning Source LLC
Chambersburg PA
CBHW021759230426
43669CB00006B/127